T0342515

The Glass City

The Glass City

Toledo and the Industry That Built It

Barbara L. Floyd

University of Michigan Press
Ann Arbor

Published in the United States of America by
The University of Michigan Press
Printed and bound by CPI Group (UK) Ltd, Croydon, CR0 4YY

2018 2017 2016 2015 4 3 2 1

A CIP catalog record for this book is available from the British Library.

Library of Congress Cataloging-in-Publication Data

Floyd, Barbara.
The glass city : Toledo and the industry that built it / Barbara L. Floyd.
pages cm
Includes bibliographical references and index.
ISBN 978-0-472-11945-5 (hardcover : acid-free paper) —
ISBN 978-0-472-12064-2 (e-book)
1. Glass manufacture—Ohio—Toledo—History. 2. Toledo (Ohio)—History. 3. Toledo (Ohio)—Economic conditions. 4. Libbey, Edward Drummond, 1854–1925. 5. Businessmen—Ohio—Toledo—Biography. 6. Toledo (Ohio)—Biography. I. Title.
TP853.O3F55 2015
338.4'766610977113—dc23

2014015521

Acknowledgments

As with any such effort, this book reflects the contributions of many. First, I would like to thank my fellow archivists who assisted me in locating collections for my research: Julie McMaster at the Toledo Museum of Art, Stephen Charter at the Center for Archival Collections at Bowling Green State University, and Tom Felt at the West Virginia Museum of American Glass. My thanks also to other scholars of Toledo glass whose research informed this work, particularly Jack Paquette, Bill Hamilton, and Stuart W. Leslie. Also, Larry Nelson and the editorial board of *Northwest Ohio History* gave me permission to use research previously published in that journal. Gary Pakulski, one of the best business writers in Toledo (now retired, I am sad to note) and a longtime friend, read an early version of the manuscript and provided some insightful comments that have made this a much better book. I thank my colleagues Arjun Sabharwal, Tamara Jones, and Sara Mouch for helping with the everyday functions of the Canaday Center while I worked on this project. Scott Ham provided a keen eye as my editor, and gave me the confidence to see the work through to the finish. And my husband, William Little, endured many weekend mornings when I would get up long before sunrise to work on my writing. I apologize for those early alarms.

But this book really reflects and celebrates the thousands of men, women, and children who have toiled doing hot, dangerous work in Toledo's glass industry for the past 125 years: the stokers of the furnace fires; the artisans who shaped the gobs of molten glass at the end of blowpipes; the young blowers' dogs who molded the bottles; the polishers, the grinders, and the cutters; the line workers who tended the automatic machines; the inventors of new products and processes; the packers and the shippers; and the countless others performing jobs big and small, both white and blue collar. While this book may dwell on the achievements of the corporate leaders, no one should forget that it is the workers who have made glass king in Toledo for the past 125 years.

Acknowledgments

Contents

Introduction

Toledo glass was used to make the spacesuits of the astronauts who landed on the moon in 1969, and it was used by Admiral Richard E. Byrd in scientific experiments he conducted at the South Pole in the 1930s. It covers the airport in Mecca in Saudi Arabia, where millions of Muslims come each year in their pilgrimage to pray, and it encased the towers of the World Trade Center. It protected America's Declaration of Independence in the National Archives, and it has been used by revolutionaries around the world to convey their beliefs with Molotov cocktails. It has held the punch served at receptions in the White House, and the alcohol in the brown bags of paupers on street corners everywhere. It insulated the Alaskan oil pipeline, and it is used in solar energy panels. It is displayed in some of the finest art museums in the world, and every day it is tossed into garbage pits. It literally surrounds us in our windows, walls, and roofs, and it holds our water as well as our wine.

But while Toledo glass has impacted the world, its most important impact has been on the city where it began. In Toledo, glass has been king since it was so declared in a headline in the *Toledo Blade* 125 years ago.[1]

Glass became king because city fathers were looking for an industry that would ensure Toledo's destiny of greatness that was proclaimed by real estate investor and newspaper publisher Jesup W. Scott in 1868. While today we may chuckle at Scott's vision of Toledo as the "Future Great City of the World," that vision of greatness led to Toledo becoming the "Glass Capital of the World." It is the vision that propelled city fathers to offer a struggling glassmaker in East Cambridge, Massachusetts, $4,000 and land in 1888 if he would bring his failing company to Toledo. That man—Edward Drummond Libbey—arguably is the most important individual in the city's history. The enormous fortune he made from the glass company that bears his name today still funds one of the premier art museums in the country, which he and his wife, the granddaughter of Jesup Scott, founded. In addition to bringing high culture to the city, that fortune has also helped generations of the young, the poor, and the disabled of Toledo.

Libbey's one company spawned three more major corporations. While the technological innovations that would produce Owens-Illinois, Libbey-Owens-Ford, and Owens Corning were the work of others, it was Libbey's success that was at the root of all. It was Libbey who discovered the genius of Michael Owens, and funded his work to develop a machine to automatically produce bottles that led eventually to Owens-Illinois. At Owens's request, Libbey also invested heavily in a machine to draw flat glass that, while not invented by Owens, was perfected by him. That effort produced Libbey-Owens-Ford. And while Owens-Illinois began experimenting with the production of Fiberglas in the 1930s—experiments that would lead to Owens Corning—it was Libbey who, 40 years before, had dazzled the country with a glass fiber dress displayed at the World's Columbian Exposition.

What has Libbey's legacy meant to Toledo? One need only take a drive through the downtown to see that the tallest buildings that dot the skyline once housed the corporate headquarters of Toledo's glass companies. From their offices on the top floors of the building on Madison Avenue, the Fiberglas Tower, and One SeaGate, the corporate leaders of Toledo glass shaped the city's physical landscape, exercised control over huge aspects of its economy, and held sway over the decisions of its political leaders. The companies brought great wealth to their leaders, but they also brought prosperity and a path to the middle class to tens of thousands of other Toledoans.

Yet the glass industry today seems more a part of the city's past than its future. While Toledo may continue to claim that it is the "Glass Capital of the World," that title is in some doubt. An article appearing in the *Wall Street Journal* in 2010 claimed that several Chinese cities more rightfully are vying for the title.[2] Today, 45 percent of all glass is produced in China, where there is less environmental regulation and cheaper labor. Even the glass that encases the award-winning Toledo Museum of Art's Glass Pavilion—a gallery that honors Toledo's glass heritage—was made in China.

Regardless of where the glass industry's future lies in Toledo, this book attempts to examine how important it has been to Toledo's past. It also attempts to put that history within the context of other historical events—to be more than just corporate history. Looking through history's rearview mirror, one cannot help but see that some decisions made by the glass corporations were not always good for the city, but it is important to understand how those decisions produced the city that exists today. If there are any lessons to be learned in examining the history of the glass industry in Toledo, it might be that like rubber in Akron, steel in Youngstown, and automobiles in Detroit, a city's economy based largely on a single industry may not be sustainable.

While the glass industry may play a much smaller role in Toledo's future, the common history between the city and the industry forever binds the two together, and it should not be forgotten.

ONE

In Search of Greatness

Cities are organisms that grow up as naturally as men. They develop where human faculties are most effective, and because these faculties can be more effective there than elsewhere.

—Jesup W. Scott, *A Presentation of Causes Tending to Fix the Position of the Future Great City of the World*, 1868

Geography was unkind to northwest Ohio.

When the glaciers retreated from the area about 12,500 years ago, the region's land was leveled nearly flat. In addition, deposits of heavy clay soil that drained poorly produced a swamp almost impenetrable by humans. As described by Dr. Daniel Drake in his 1850 book *A Systematic Treatise, Historical, Etiological, and Practical, of the Principal Diseases of the Interior Valley of North America*, "Between the Maumee and Sandusky Rivers, south of the western extremity of Lake Erie, lies the great forest, which has received the ominous name of the Black Swamp. . . . The levelness of this tract, taken in connection with the argillaceous bottom, explains the paludal or swampy character of its surface. From this surface arises a miscellaneous forest, of greater density and loftiness than is to be found elsewhere, perhaps, in the Interior Valley of North America."[1]

The geography of northwest Ohio meant that while the rest of the state experienced considerable development in the years after statehood in 1803, northwest Ohio lay largely dormant. The swampy conditions made it difficult to move through the territory except by navigating its streams and the Maumee River. The conditions also made it unhealthy for the few who did. Cholera, malaria, typhoid, leptospirosis (called "autumnal fever" by Dr.

Drake), and smallpox regularly swept through in epidemics, earning Toledo the nickname "Grave of the Midwest."[2]

In addition to geography, the Native American population and their allies also had stunted white settlement of the area for many years. The French, who claimed northwest Ohio based upon exploration in the sixteenth century by Jacques Cartier, established alliances with the Native Americans in the hope of keeping out the English. The English, many of whom arrived from Virginia and Pennsylvania over the Allegheny Mountains, claimed the region based on the exploration of John Cabot in 1498. By the beginning of the eighteenth century, the Maumee River was already important in the fight between France and England over control of the region in an ongoing war that came to be known by the British as the French and Indian War, lasting from 1754 to 1763.

The primary Native American tribes in northwest Ohio were the Ottawas and Wyandots. Others who settled in areas around the Maumee Valley included the Shawnee, Delaware, and Miami tribes. The name "Me-au-me" was given to the region's major river by French explorers who understood this to be how the Miami tribe said its name. The Native Americans mostly sided with the French in the war with the British because the French provided them with supplies, while the British withheld them as a way to exert control.

Pontiac, who became a leader of the Ottawa tribe, was born in northwest Ohio and lived most of his life in the Maumee Valley. In 1763, he planned an all-out attack on British settlers in the region in an effort to wipe out the entire population and retain control by the tribe's French allies. The "conspiracy," as it was called, resulted in the death of many settlers as well as Native Americans in a brutal period of fighting. Pontiac eventually gave up when he discovered that the French had already ceded the territory to the British in a peace treaty and had retreated to Canada. But other Native Americans did not give up, and in 1772 joined together in a confederacy of the Shawnee, Wyandot, Miami, Ottawa, and Delaware tribes. They continued their campaigns against the British settlers.

The Treaty of Paris that ended the Revolutionary War in 1783 was supposed to force the British out of the Maumee Valley and open up the area for settlement by citizens of the new United States. But the British stubbornly refused to leave several of their fortifications in the region, including Detroit. After the defeat of the French, the Native American tribes switched their allegiance to the British in hopes of keeping the Americans out of the area.

The Northwest Ordinance of 1787 established U.S. government control over the vast areas around the Great Lakes. The country did not want to be distracted by Indian uprisings in the region, but despite treaties signed with some tribes, attacks on and by both sides continued. General Josiah Harmar and General Arthur St. Clair each tried to end the Native American problem by crushing the tribes in 1790 and 1791, but were unsuccessful. St. Clair's expedition was particularly disastrous for the Americans, as he lost nearly 600 men. In 1793, the British built a new fort on the Maumee River called Fort Miami at a location 13 miles from present-day Toledo, and the Native Americans began encampments near the fort because they believed the British would provide them with protection in case of attack by the Americans.

After the disasters of Harmar and St. Clair, President George Washington appointed General Anthony Wayne to lead a campaign aimed at finally settling the Native American problem in the Maumee Valley. On August 20, 1794, Wayne met a confederated force of Delaware, Miami, Ottawa, Iroquois, Chippewa, and Pottawatomi Indians at an area where many trees had been felled by a tornado the year before. The Battle of Fallen Timbers was a complete victory for Wayne, despite being outnumbered two to one. Wayne lost just 33 soldiers in contrast to hundreds of casualties for the Native Americans. The Native Americans retreated to Fort Miami (about five miles away), where they expected to be offered safety, but were turned away by the British. With the overwhelming defeat, the Native Americans met Wayne in August 1795 and signed the Treaty of Greenville. The treaty gave the Native Americans parts of northwestern Ohio while establishing several military reserves in the region. The British continued to agitate with the Native Americans against the American settlers, causing problems for the United States in the region and ultimately contributing to the cause of yet another war with Great Britain.

"No section of the country was made the scene of or was called to suffer more severely the unfortunate consequences of the War of 1812–1815 between the United States and Great Britain than was the Maumee Valley," historian Clark Waggoner stated.[3] After General William Hull surrendered Detroit to the British with little fight in August 1812, northwest Ohio was left without any defense. Some settlers lost their property to all three forces fighting in the war—the British, the Indians, and the Americans. Two particularly brutal assaults killed many. At what became known at the River Raisin Massacre, near Monroe, Michigan, U.S. troops under General James Winchester were caught unprepared by Colonel Henry Proctor, and most were killed or captured. At Dudley's Massacre in Maumee, over 600 men under the command of Colonel William Dudley were killed or captured by Colonel Proctor when they were ambushed while trying to disable British cannons at Fort Miami. The success of General William Henry Harrison in outlasting two sieges by the British under Proctor and the Native Americans under Tecumseh at Fort Meigs near Perrysburg in 1813 and the defeat of the British on the Great Lakes by Commodore Oliver Hazard Perry were pivotal in removing the British and their Native American allies permanently from the Maumee Valley—and in winning the war for the Americans.

The Founding of Toledo

Despite the difficulties presented by the swampy terrain, the end of the War of 1812 led land speculators to lay out a town called Port Lawrence on the Maumee River at the mouth of Swan Creek in 1817. But unable to attract many buyers, the speculators defaulted on their loan the following year and the land was returned to the government. Many of the settlers left, but some, including Joseph Prentice and John Baldwin, stayed and established a warehouse and a lake shipping business, laying the foundation for the settlement's early economy. After the failure of the first Port Lawrence, a second settlement by the same name was plotted in 1832.

But Benjamin F. Stickney, who had settled in Port Lawrence, felt progress in the town was too slow, and in 1832 he withdrew his interest and created an adjacent new town called Vistula. Working with Edward Bissell from New York, the two began improving their town with buildings and roads. While speculators attempted other settlements along the river, most quickly failed, leaving Port Lawrence and Vistula as rivals for development.

The towns remained rivals until plans to build a terminus somewhere on the Maumee River for a new canal system that would link up the west with the Erie Canal in the east were announced. The leaders of Port Lawrence and Vistula decided to consider merging in 1837 in order to strengthen their bid for the canal against competitors Maumee and Perrysburg.

While the citizens of the two towns agreed that a merger was in their best interest, they could not agree on a name for their new entity. Exactly how the settlement came to be called "Toledo" is unclear. Historian Waggoner claimed it was James Irvine Browne, at a joint meeting of the residents of Port Lawrence and Vistula, who suggested the name based on his travels throughout Europe.[4] Historian Harvey Scribner conjured that it might have been Two Stickney, son of Benjamin Stickney, who, while examining a map of Spain, came across the name Toledo and thought it would be a good one.[5] S. S. Knabenshue, in a column that appeared in the *Toledo Blade* in 1903, said the name was suggested by Willard J. Daniels, a merchant in Vistula and a landowner in Port Lawrence, who had been reading Spanish history and liked the name. According to Knabenshue, "His [Daniels's] argument was that the word was easily pronounced, is pleasant in sound, and that there was no place of that name on the western continent."[6] Regardless of who suggested the name, Spanish culture was much in vogue in the United States at the time, and the name was a direct homage to the town in Spain.

After years of being neglected by developers, plans for a northern outlet for what was to be called the Miami and Erie Canal finally spurred an interest in Toledo. But before the canal could be built, it had to be decided whether Toledo was located in the state of Ohio or the territory of Michigan. The dispute over the boundary was the result of different maps of the region used in drafting the Ordinance of 1787 and the enabling act that created Ohio in 1803. The northern boundary of Ohio was supposed to exist along a line drawn from the bottom of Lake Michigan directly east. But where that line actually existed was disputed by Ohio and Michigan. Because of the potential for economic development in the region with the construction of the canal, Ohio moved quickly to secure the territory.[7]

Governor Robert Lucas asked the Ohio's legislature to exert the state's control, which angered Michigan's territorial leaders. It was claimed by Michigan that most settlers in the area preferred to be a part of the Michigan territory. In March 1835, General Joseph Brown, commander of the Third Division of the Michigan Militia, rallied his troops by stating, "We cannot submit to invasion of our soil. We are determined to repel with force whatever strength the State of Ohio may attempt to bring into our Territory to sustain her usurpation, and let the consequences which may follow, rest on the heads of those who attempt to deprive us by force of our rightful jurisdiction."[8] Ohio countered by creating Lucas County (named for the Ohio governor), which included the disputed territory.

The rhetoric between the state and the territory continued to escalate. Surveyors sent out to determine the correct boundary encountered Michigan militia forces, who harassed them. Ohio sympathizers were arrested. Scuffles continued, with Two Stickney arrested for stabbing and injuring a Monroe County (Michigan) deputy sheriff. President Andrew Jackson sent emissaries to cool both sides and ordered that the survey continue, removing Michigan governor Stevens T. Mason from his position because of the agitation.

To establish civil control over Toledo, Governor Lucas held a Court of Common Pleas session September 7, 1835, shortly after midnight, despite 500 to 600 Michigan volunteers on their way to attempt to stop the court from meeting. Because the court was held in the middle of the night, Michigan claimed the appointment of county officials by the judges was illegal. But the court session cemented Ohio's control of the area, and the soldiers under former governor Mason's control went home peacefully. The Toledo War ended officially in 1836 with the passage of the Clayton Act by Congress.[9] The boundary line was drawn to include Toledo within the state of Ohio. The real loser in the war was Wisconsin. The territory's northern portion was taken away and awarded to Michigan in exchange for Toledo going to Ohio.

A poem written by an anonymous author known only as "Bard of the Woods" in March 1835 seemed to sum up the irony of a war fought over a place that no one had wanted for many years:

> Young Toledo! Rise to Fame!
> Mart of the Western World should claim
> Homage to all the ports around—
> Her wealth and power know no bound;
> More mighty far than ancient Rome,
> Stand by inherent power alone
> But oh! Methinks I see them dashing!
> Hear pistols pop! And swords clashing!
> While first to last many oppose,
> With eyes plucked out or bloody nose;
> Whose horrid threatening or grimace
> Convince they'll die or keep their place.
> The first of April is the day,
> For Ohio bravos to display.
> According to all ancient rules,
> No doubt they'll all be "April Fools."[10]

Toledo's Struggles

Despite Toledo's founding and "victory" in the war over its placement on the map, the settlement continued to struggle as it had for decades. The biggest problem was its swampy surroundings. The Black Swamp made travel from

Jesup W. Scott, author of *The Future Great City of the World.* (University of Toledo Archives.)

the east into the region difficult along the only road, appropriately called Mud-Pike. The road was impassable until 1838, when $40,000 was allocated for paving what is today State Route 20. In 1833, Jesup Scott, who owned the land where today the Lucas County courthouse is located, got lost in the swamp there and could not find his way out. In 1839, city council authorized the street commissioner to take measures to drain and improve the swampy areas of the city, and the first sewers were constructed in 1848. But flooding of the Maumee River was common, and the new sewers actually made the problem worse by quickly channeling the water into the river during downpours. Ice dams in the winter also exacerbated flooding. The city was nicknamed "Frog Town" because of the large numbers of the amphibians living in the city.

But the building of the canals turned Toledo into a hotbed of speculative land investment. Talk of a canal connecting Lake Erie to the Ohio River began as early as 1819. In 1824, plans for a canal connecting Cincinnati and the Ohio River to Maumee and Lake Erie were approved, but construction was delayed several times.[11] Other plans called for a canal linking Toledo with the Wabash River to the west. Construction delays did not deter those who sought to make their fortunes buying and selling land. As Jesup Scott noted, "In 1835, commenced that memorable speculation in wild lands and wild cities, which culminated in 1836. The whole Maumee valley was filled with eastern fortune hunters."[12]

The Wabash and Erie Canal opened in 1843 and connected Toledo with Lafayette, Indiana. The 323 miles of the Miami and Erie Canal finally opened in 1845, and connected the city to Cincinnati. The success of the canals established Toledo as the dominant settlement in northwest Ohio, leaving behind its rivals of Maumee and Perrysburg. In 1848, the number of canal boats traveling through the city numbered nearly 4,000.[13] In 1851, the total value of exports from Toledo via the canals was nearly $8 million, while Maumee and Perrysburg handled less than $75,000 of exports that year. Statistics on imports was even more impressive, with Toledo importing over $23 million in goods compared to just over $275,000 for Maumee and Perrysburg combined.

At nearly the same time the canals were being completed, railroad lines were under construction in northwest Ohio. In the end, the railroads would dominate and the canals would quickly fade. The first railroad was the Erie and Kalamazoo Railroad, established in 1835, which featured cars drawn by horses. There were many competing railroads, and eventually in 1856 the lines came together to create the Toledo, Wabash, and Western Railway. By 1870, the success of the railroads led to the abandonment of the canals, which soon blighted the downtown as deserted boats rotted and filthy water flowed through the city. But the burgeoning economy produced by both the canals and the railroads led to a doubling in Toledo's population between 1850 and 1860, with the population of Lucas County increasing from 25,831 to 46,722.[14]

Despite a growing economy and population, living in Toledo continued to be difficult at best. Cholera outbreaks were a constant problem, with epidemics occurring in 1849, 1852, and 1854. The *Toledo Blade* reported in 1854 that 239 residents died in 17 days in July alone.[15] But the paper also argued that the cholera alarm was overstated, as most of the sick "had violated all the laws by which human beings exist. Many of them are transitory people, evidently vagrants and vagabonds who happened to tarry here long enough to die." "Ignorance, neglect of symptoms, intemperance and filth, are the cause of nine-tenths of the disease," the paper reported. Toledo was without a hospital, an orphanage, or nursing care during the epidemics. The Reverend Augustine Simeon Campion, pastor of St. Francis de Sales Church, traveled from Toledo to Montreal, Canada, to ask members of the Sisters of Charity order to establish an orphanage in the city to care for children whose parents had died of cholera. In late October 1855, four Sisters arrived in Toledo and founded the St. Vincent Asylum, which would evolve into the city's first hospital.

The number of businesses and professionals in the city continued to grow despite the bleak living conditions. In 1835, it was still possible to list each individual business in the city.[16] By 1858, the numbers were too great to list each business individually, and only a summary of the number of types of business was listed. Included in the listing were 25 attorneys, 30 boardinghouses, 66 grocers, three daguerreotype galleries, four newspapers, and one portrait painter.

The Civil War as Watershed

As in many northern cities, the Civil War was a watershed for Toledo's development as an industrial city. With southern ports no longer available to ship goods into the interior, and demand high for all of the products needed to wage war, Toledo's economy took off. The total capital invested in all businesses in the city in 1860 was $660,700. By 1880, that number had increased to $5.5 million.[17]

On April 15, 1861, following the attack on Fort Sumter, citizens gathered in downtown Toledo in unity in their opposition to the South. They passed a resolution that read, in part, "Resolved, that as citizens, we pledge ourselves to ignore all past party distinctions and give our united aid and support to our Government; to protect the Capital; maintain the Government; punish the insult offered our Flag; and restore peace and tranquility to the Country."[18] The first company of volunteers for the Union army from Lucas County were sworn in two days later.

Toledoans served bravely and sacrificed much in the war, with 293 men from Lucas County reported killed. The city also produced two important heroes—one on the battlefield, and one in Congress. From 1857 to 1861, James B. Steedman was editor of the *Toledo Times* and a lawyer. As a colonel in the Union army, he was the hero of the battle of Chickamauga, playing a vital role in saving the battle for the North. Toledo's Congressman James M. Ashley introduced a constitutional amendment in 1863—the first of its kind—calling for the abolishment of slavery. Ashley's amendment would finally be passed in 1865 as the Thirteenth Amendment to the U.S. Constitution.

The Future Great City

Jesup W. Scott was what might be called a "geographic determinist." As early as 1828, he began studying where he thought areas of vibrant economic development might be located in the young United States. He first moved to northwest Ohio in 1832, and he invested in the real estate boom that was occurring at the time as plans developed for canals in the city, buying 70 acres of land in Toledo. He became wealthy from his land deals, and at one time owned much of what is today downtown Toledo. After briefly returning to his native Connecticut, he moved back to northwest Ohio in 1844. He served as editor of the *Toledo Blade* until 1847. He also contributed articles to national periodicals such as *Hunt's Merchant Magazine* and *DeBow's Review*, which were influential in shaping intellectual discourse in nineteenth-century America. Many of these articles addressed his theories concerning geography and economic development.

In 1868, he expanded on his ideas in a pamphlet entitled "A Presentation of Causes Tending to Fix the Position of the Future Great City of the World in the Central Plain of North America."[19] Here, Scott laid out his theory that since ancient times, the world's economic center had been moving westward. Most recently, he pointed out, the location of the world's largest city had shifted from London to New York. He believed that it was only a matter of time before the next great commercial center would be located in the interior regions of North America—and perhaps not surprisingly for someone who owned real estate investments there—this would likely be Toledo.

For Scott, cities were like organisms, and they sprouted up in areas where humans could be most effective.[20] He believed that climate played an important role in determining growth. Scott traced the location of the great cities that had existed up to that time, and found that all were located within the same climatic latitude. The reason for this, he believed, was that this temperate climate made men vigorous. Also important to a city's development were natural waterways and interior commercial routes. The Great Lakes provided both—cool temperate breezes and navigable waters. Furthermore, he stated that an interior city would become the world's largest because most future commerce in the United States would be with interior markets, not foreign ones where coastal cities would have the advantage.

In Scott's view, the next commercial center would likely be either Toledo or Chicago. But he believed Toledo would eventually surpass Chicago in growth once the dense forests surrounding Toledo were felled and it, too, was surrounded by prairies like Chicago. The dense forests would also supply needed lumber for Toledo's growth. Scott felt that within 100 years, Toledo would realize its full potential. "The cities of western Europe are grand outgrowths of modern improvements, but they will be deemed, in their present condition, rude and small in comparison with the vast emporiums which, in one hundred years, will grow up on our continent," Scott concluded his pamphlet.[21] To help Toledo fulfill its certain fate, in 1872 Scott endowed a university to train "artists and artizans" for their role in the Future Great City. The Toledo University of Arts and Trades was a failure, but it was resurrected by Scott's sons as a manual training school. It would later evolve into the University of Toledo.[22]

Scott's boosterism helped to fuel speculation among many Toledo leaders that the city was just one step away from greatness. Industrialists from other areas of the country began moving to the city. The Milburn Wagon Works moved to Toledo from Mishawaka, Indiana, in 1873, and became the largest manufacturer of farm wagons in the nation and expanded production to include buggies and delivery wagons. Peter Gendron started a company in Toledo in 1877 that produced wire-spoked wheels that were used in many types of vehicles, including wheelchairs, children's toys, and bicycles. Bicycles were a craze in the late nineteenth century because they were better suited for the urban environment than polluting horses. They were also easy to maintain and easy to store. Toledo played a major role in manufacturing bicycles, and by 1890 was the largest producer in the country. In addition to Gendron, other bicycle manufacturers included the Kirk Manufacturing Company, Lozier and Yost, and Union Manufacturing. In the early twentieth century, many of these wagon and bicycle companies evolved into early automobile manufacturers. The R. L. Polk and Company directory of 1887 for Toledo contained 12.5 times more business names than those listed just 20 years before.

Despite these successes, many Toledo leaders were still waiting for industries that could truly realize the city's great potential as outlined by Jesup Scott. In order to attract such industry, the city would need to be able to provide cheap natural gas. Their dreams seemed near when, in 1884, amateur geologist Dr. Charles Oesterlin discovered pockets of high-pressured natural gas in Findlay, Ohio, just 50 miles south of Toledo. Within two years, 17 wells had been successfully drilled around Findlay, and the search was on for more in adjacent counties. The price of land near successful wells soared from $50 an acre to between $500 and $1,000 an acre.[23]

Surely similar deposits had to exist under Toledo. "Each new well discovered seemed to be bringing these treasures of nature closer to her borders, and to offer greater hopes that the drill would soon prove gas was within her limits," Clark Waggoner noted in his history of the city, which was published while the hunt for these deposits was still ongoing.[24] Many thought natural gas was self-generating and therefore inexhaustible; thus the more that was used, the more that was created.

The gas hysteria led to many unsuccessful efforts to find deposits under Toledo. In 1885, banker Horace S. Walbridge proposed a test well that would be paid for by the city—which would quickly recoup the cost when gas deposits were found. But only a small amount was discovered before the drill hit limestone and flint rock. In 1886, the Citizens Natural Gas Company of Toledo was formed to try another test well, but it too was unsuccessful in finding large deposits.

There was fear that smaller towns around Toledo that had successfully found gas and oil would soon surpass the city in development. For the city fathers, this was an insult to Toledo's clearly bright and certain future. As an article in the *Daily Blade* in February 1887 noted,

> Toledo to the front.
> There is no time to lose.
> Other cities are piping oil and gas.
> Small towns in Northwestern Ohio are booming.
> Toledo, with oil and gas, must share the boom.
> Toledo has unrivaled commercial facilities.
> It is the railroad center of the West.
> It is the spot where iron, coal, lumber and grain will meet.
> It is destined to become the manufacturing center of the West.
> And not only the manufacturing and commercial center, but the future oil and gas center of Ohio.[25]

The first successful natural gas well inside the city was drilled in April 1887 in the Manhattan area near Point Place, but most others produced nothing. After repeated failures to find gas under the city, the Toledo Natural Gas Company was formed to construct a pipeline from successful fields outside the city and bring the gas to Toledo. It attracted investors from as far away as Pennsylvania. A competing company, the Northwestern Ohio Natural Gas Company, also sought to build a gas pipeline. The race was on.

Despite little tangible evidence of large deposits being close at hand, city leaders were not beyond promoting easy access to natural gas to manufacturers who might be attracted to cheap, abundant fuel. The Toledo Business Men's Committee formed on April 8, 1887, to promote Toledo's industrial prospects. The mission of the organization was to present "a united, systematic effort to improve the unequaled advantage Toledo possesses above all her competitors."[26] The group began to heavily promote the city, running advertisements in newspapers throughout the country, but especially in the East.

An article about the group in the May 5, 1887, *Daily Blade* referenced Jesup Scott's treatise, boasting in a headline, "'Future Great' No Longer. Toledo has become the 'Present Great.'"[27] The article went on to say that "Toledo needs to-day more men—more men of enterprise; more men with brains; more men with capital; more men who are skilled employers. There are fortunes to be made here in the next few years. Money will be poured out at the feet of men who will build pipelines for gas and oil, plant iron and steel manufacturers, start glass factories, local oil refineries, run woolen mills, erect mills for the manufacture of woodenware, and give the city what is needed—more manufactories."[28] The article even turned Toledo's arrested development into a strength. "The growth of Toledo is without parallel in the history of the New World. Seventy years ago a wilderness, now the garden spot of the Great Northwest. Only 70 years!" the *Daily Blade* proclaimed, as if it has happened overnight.[29] Toledo was not a boomtown, but rather like a "sturdy oak" that had taken years to mature.

In July 1887, the Northwestern Ohio Natural Gas Company completed its pipeline into the city, beating out its competitor by a few hours. The completion of the pipelines fueled another Toledo real estate boom, with many investors from outside the area buying up land in the city. Contributing to the frenzy was news in the local papers that hinted to negotiations that were under way with "huge" factories—including glass factories—to relocate to the city.

On September 7, 1887, the city held a gas celebration. Natural gas stand pipes all around the city were lit, and residents and guests invited for the event rushed out into the streets to see. The fires could be seen for miles. Former president Rutherford B. Hayes and former congressman James Ashley spoke to an assembly at Memorial Hall to mark the occasion. President Hayes read from a letter from Edward Orton, state geologist of Ohio, who expressed his belief that the natural gas supplies under the state would endure longer than those elsewhere. But he also urged that the gas not be wasted, an ironic statement given the events of the day. The *Daily Blade* described the day's event as another expression of Toledo's clear destiny to greatness. "By burning jets of natural gas, by brilliant flames from mammoth gushers, by the roar from the combined force of forty wells, thousands of strangers and citizens read last night the destiny of Toledo. Toledo, the queen city of the lake, goes forth conquering, one hand bearing a torch with light for the world, with fire for a nation's forges, with heat for a million looms, with fuel for thousands of factories; the other holding cheap iron, lumber, copper, oil, wheat and wool."[30]

Future Great City of the Universe?

The campaign of the Toledo Business Men's Committee to get companies to relocate to the city targeted glass manufacturers in particular. At the time, natural gas was not used much in the glass industry, with the Rochester glassworks in Pittsburgh being the first to use it in 1880. But many glassmakers believed natural gas was superior to coal in fueling glass ovens. The committee called attention not only to the natural gas available in Toledo, but also noted the city's access to oil, its established transportation lines, and its location on a stratum of quartzite that ran from Sylvania in the west to the Maumee River in the east. All were necessary for glass factories, and all were waiting to be successfully exploited.

Even after Jesup Scott's death in 1875, his vision of Toledo as the "Future Great City of the World" continued to have an enormous impact on the city. Every test drill for gas, every acre of land bought in speculation, every new industry attracted to the city—all supported Scott's argument. Clark Waggoner's 1888 history of the city is full of fervor about Toledo's future. In addition to statistics that all seemed to point to unbridled growth, the volume also contained biographies of leaders who were helping to create the city.[31] Few were born in Toledo—rather they moved to the city because they believed their fortunes could be made here. Like Scott, they awaited Toledo's greatness.

In January 1888, the *Daily Blade* carried a letter to the editor that dared to predict what Toledo would look like 111 years in the future. Entitled "A Garden of Eden," it predicted that by 1999 the city would have 6 million residents, and would be rivaled in size only by New York City. Detroit would be a northern suburb, and portions of Canada would be annexed to Toledo.

These predictions for Toledo in 1999 were only the beginning, however. "But Toledo, the world's metropolis, is not satisfied yet. The question which now agitates her is whether an inter-planetary air line cannot be established with our nearest neighbors, Mars and Venus," the article noted, only half in jest.[32]

TWO

Boss-Town

From Boston to Boss-Town.
From the Hub of the East to the Hub of the West.
From the City by the Sea to the City of the Lakes.
From the Land of Classical Culture to the Land of Natural Gas.
All Toledo welcomes you to the future glass center of the world.

—From an article in the Toledo *Daily Blade*, August 18, 1888, about the arrival of Edward Drummond Libbey and his workers to Toledo

Glassmaking was the first industry in the American colonies. In 1608, the London Company sent Dutch and Polish glassworkers to Jamestown to teach the colonists how to make glass, and a glass factory was established at the struggling settlement. The factory was to make vessels and window glass for the colonists to use and to trade. It was unsuccessful. In 1621, a second group of glassmakers was sent to the settlement to make glass beads to trade with the Native Americans. But the Native Americans revolted against the colonists the following year, and the glass industry, like the original colonial settlement, failed to survive.[1]

Other early colonial settlements also established glass furnaces, including Salem, Massachusetts, in the 1640s and Philadelphia in the 1680s. But most glass used in the colonies was produced in England, where glass was a well-established industry. Venetian glass artisans, whose skills were known around the world, migrated to England beginning in the sixteenth and early seventeenth centuries and manned the British factories. After the American Revolution, there was limited domestic development of glass manufacturing in the young United States because Europe dumped its cheap glass the new country. A small American glass industry grew during the War of 1812 when a shipping embargo cut off foreign supplies. But after the war, Europe returned to dumping glassware on the country, making a competitive domestic industry difficult. Because of the lack of trade protections—which glassmakers frequently complained about—few fortunes were made in American glass before the 1860s.[2]

Domestic glass production finally became viable because the Civil War again disrupted foreign imports, high protective tariffs were finally enacted, and the raw ingredients needed for glass production were found to be plentiful, cheap, and of high quality in the United States. Between 1820 and 1880, the number of glass furnaces in operation grew five times, and the number of workers grew 25 times.[3] But while the industry expanded, it suffered from a shortage of skilled workers with the knowledge to make glass. The best glassmakers still learned their craft in Europe, and few skilled glassmakers emigrated to the United States.

American glassmakers made all the necessary products—window glass, bottles, and tableware. But production methods remained stuck in time, with little innovation. This was partly due to the enactment of high tariffs and the resulting lack of foreign competition, which led to poorer quality domestic glass. The techniques of glassmaking in nineteenth-century America differed little from those of the Jamestown settlement of the seventeenth century.

The New England Glass Company

The earliest commercially viable glass production in the new United States was centered along the east coast, mostly in Massachusetts. The first successful glass factory was the Boston Crown Glass Company, founded in 1787.[4] It made window glass. In 1813, the Boston Porcelain and Glass Manufacturing Company built a factory in East Cambridge, a suburb of Boston. It hoped to cash in on the demand for glass resulting from the embargo of the War of 1812. It made both crown glass for windows and high-end flint glass. But before it could establish a strong domestic market, the war ended, and with it, the embargo that had fostered its development. A second attempt to make the factory successful in 1815 also failed.

Although little is known about him, Deming Jarves was one of the first successful American glassmakers.[5] He began his career producing crockery ware in 1815 for a firm that bore his name, Henshaw & Jarves. In 1817, at the age of 27, he bought the plant and property of the former Boston Porcelain and Glass Company in East Cambridge, and in 1818 established the New England Glass Company. The growing economic prosperity of the United States produced a market for the richly colored cut glass that Jarves produced. The six-pot furnace at the New England Glass Company employed 40 skilled and unskilled glass laborers and produced $40,000 in goods its first year.[6] By the next year, its output had grown to $65,000 in products. In addition to cut glass, the company also introduced pressed glass in the 1820s, the first factory in the country to produce such glassware.[7]

Jarves left the company in 1825 over differences with the other directors to create a new company, the Boston and Sandwich Glass Company, which became the chief competitor of the New England Glass Company. In 1826, the skilled glassworker Thomas Leighton became superintendent of New England Glass, and under his direction sales of the company's cut glass reached $150,000 a year.[8] The company's fine cut glass was highly sought after, and it established sales offices in New York, Philadelphia, and Baltimore. By 1851, the company was worth over $400,000, employed 450 workers, and was said to be the largest glass factory in the world.[9]

The success of the New England Glass Company encouraged others to enter the glass business. In 1860, Timothy Howe and William L. Libbey founded the Mount Washington Glass Works in South Boston.[10] Libbey had learned the business side of the glass industry working as a clerk for Deming Jarves in 1850. In 1866, Libbey took over sole ownership of Mount Washington after Howe's death, and in 1869 moved the company to New Bedford, Massachusetts. Around 1872, Libbey left his own company to become an agent for the New England Glass Company.

The New England Glass Company continued to make flint glass exclusively. Flint glass differed from other glass in that it used red lead and only the highest quality of other ingredients like white sand and potash. It was heavy, and when you struck the side of it, it produced a metallic-sounding ring. But in 1864, William Leighton, son of Thomas Leighton, who was working for the Hobbs & Brockunier glass company in Wheeling, West Virginia, invented the process of making lime glass. While considered of poorer quality than flint glass, it was cheaper to produce. The new factories being established in places like western Pennsylvania and West Virginia also had the advantage of being closer to the coalfields that fueled glass furnaces, and were thus cheaper to operate.

The New England Glass Company refused to lower its standards, and continued to produce high-quality leaded flint glass.[11] The company also excelled at cut and etched glass. Cut glass was made by highly skilled artisans who carefully and painstakingly cut the glass to create facets that brilliantly reflected light in a manner similar to the work done by diamond cutters. Etched glass also demanded skilled craftsmen, and it utilized hydrofluoric acid to produce beautifully detailed patterns and scenes on the glassware. The design and engraving work at the New England Glass Company was overseen by Louis F. Vaupel, an immigrant from Germany.[12] The quality of New England glass garnered much attention at the 1876 Centennial Exposition held in Philadelphia. The company regularly won awards for the quality of its products.

But the company had to balance quality against the need to be commercially successful. Not only was it competing with cheaper made domestic glassware, but also with foreign glassmakers. And the cost of shipping heavy coal to Boston made it difficult to be competitive against the increasing number of glass companies springing up near the coalfields. The cost of fuel alone accounted for a third of the price of glassware.[13] Some of the new companies in the west had also switched their fuel from coal to the plentiful and cheap natural gas and oil that was also located nearby to them.

The depression of 1873 had a devastating effect on the New England Glass Company. Few people could afford the high-quality, expensive cut glass produced by the factory. In 1875, shareholders accused the directors of mismanagement. Workers at the factory found it profitable to steal glassware and sell it themselves. By 1877, the company—once one of the largest and most profitable glass companies in the world—was on the verge of bankruptcy.[14]

With few options, the directors decided to lease the New England Glass Company factory to William Libbey. To help manage the company, Libbey brought in his 20-year-old son, Edward Drummond Libbey, as a clerk. In 1880, Libbey made his son a full partner in the company, and changed the name to the New England Glass Works, Wm. L. Libbey & Son, Props. But problems continued, and the company was operating at less than full capacity, with three furnaces idle.

Caught between the high cost of fuel and competitors making cheaper quality products, Libbey's company also faced another rising cost—labor. Glassblowers were well paid, making $9–$10 per day, or two to three times the wages of other industrial workers in the 1870s and 1880s.[15] They did not consider themselves manual laborers, but rather skilled artisans, and they were in short supply. When the workers became angry with management, or got better offers from competitors, they left. Glassworkers frequently moved from factory to factory.

In 1878, glassworkers founded the American Flint Glass Workers Union

(AFGWU). Because of its strong position, the union negotiated profitable contracts that tied the hands of the factory owners. The contacts outlined production quotas and prices. When the union organized the workers in Pennsylvania, West Virginia, and other factories in the west, it began to demand equal pay for all workers no matter their location. Those factories located along the east coast found it increasingly difficult to compete with factories using cheaper fuel when they had to pay the same wages. The eastern factories were the site of many strikes by AFGWU.

It was against this backdrop that the New England Glass Company struggled. The stress of the business clearly took a toll on William Libbey. On August 30, 1883, Edward Drummond Libbey recorded a sad entry in his batch book—the small, private book where glassmakers wrote down the formulas for their various kinds of glass. "My Dear & respected Father died upon this day at 7 a.m. It now remains for me E D Libbey to take up the finished work of my respected Parent & carry on a business left by him, in which only God alone knows how he suffered to make it a success. His life is my best legacy."[16]

The Son Continues

Edward Drummond Libbey was born in Chelsea, Massachusetts, on April 17, 1854. He studied at the Kents Hill Academy in Maine, where he received a classical education. He wanted to become a minister. His future as the head of what would become a glass empire resulted not from some personal desire, but from a sense of duty to his father. He was never a glassmaker, but rather a manager of the business side of glass companies.

Libbey's task of saving his father's company in 1883 was daunting. He was able to pay off some of the company's debt with proceeds from his father's insurance policy. But the company's real savior came from a product line of glassware called Amberina.

Amberina glass was made by adding a small amount of gold into the formula for New England Glass Company's ruby red glass batch. One story said that the unique coloration was developed by accident when a glassmaker's ring dropped into the batch.[17] The process produced glassware that varied in coloration from gold to red—all in a single piece. Libbey took samples of the Amberina glass to Tiffany's in New York, which purchased the entire stock that New England had on hand.

Amberina quickly became popular, with sales rocketing from $45,000 in

1883 to $80,000 in just one year.[18] But as quickly as it had become popular, it faded. The New England Glass Company followed the success of Amberina with other unique wares with distinct colorations, including Pomona, Peachblow, and Agata. With each new line, Libbey tried to compete with the cheaper products being produced by western glassmakers.

In 1886, the AFGWU organized the New England Glass Company workers, and began a series of painful and expensive strikes against the company.[19] The union continued to demand that workers at the company be paid wages equal to those glassworkers in plants in Pennsylvania and West Virginia. But Libbey's fuel costs as compared to his western competition made it difficult to meet the union's demands. When the company was struck again in 1888, Libbey urged the workers to end the strike, and threatened to close the plant.

The year before, the Toledo Business Men's Committee had begun placing advertisements in eastern newspapers and distributing circulars touting Toledo's natural gas boom (which was more hype than reality). The ads claimed the city offered glass factories not only the advantage of access to the plentiful natural gas located directly under the city, but also that the city's gas companies controlled large shares of "the best gas lands in Ohio" outside of Toledo.[20] The ads pointed to the ample supplies of sand, quartz, and lumber located nearby—all ingredients that were necessary for glass—and the city's location on the Great Lakes, which would reduce the cost of shipping. William H. Maher, secretary of the Business Men's Committee, offered to meet with glass company owners to answer their questions about the city's wealth of inducements.

Several glass factories expressed interest. The *Daily Blade* reported in April 1887 that a Covington, Kentucky, factory was interested in relocating in Toledo. "For the last week or two, the city has been overrun with glass men who have admired her glass sand, her transportation facilities, her location, and the hundred other excellent points," the paper reported.[21] The first hint of Edward Drummond Libbey's interest in possibly relocating to the city was reported in September of that year. While not mentioning the firm by name, the *Daily Blade* noted that a large glass factory employing 125 to 200 skilled laborers was in negotiations to secure land for a factory in Toledo.[22] Citizens were urged to talk up the city and lend their voice to aid in attracting the company. In December, the newspaper reported that Maher had gone to Boston to meet with Libbey and tour the New England Glass Company plant to assess Libbey's needs for a factory in Toledo. The paper added, "It is understood that matters have progressed so far as to make the probability of removal [from East Cambridge to Toledo] almost a certainty."[23]

By January 1888, Libbey and the New England Glass Company were in serious negotiations with the Toledo businessmen for four acres of land for a factory and 50 lots upon which to build worker housing. The housing lots had to be within a mile of the factory. The Toledo Business Men's Committee sought donations of land and money to meet Libbey's need. It was estimated that the land would cost $4,000. The firm was expected to employ 250 men. The factory buildings would be paid for by Libbey, and were estimated to cost $60,000.[24]

But negotiations did not run smoothly. By the end of January, Libbey was threatening to pull out of the deal and go elsewhere unless Toledo could meet his demands, including cheap fuel contracts.[25] Maher put out a desperate call to the businessmen of the city for support, and by February they had been successful in securing the land and money from donors. A signed agreement was reported in the newspapers on February 7, 1888.[26]

Work began immediately on construction of the factory, which was located on Buckeye Street near Ash Street in a development named Glassboro. Examples of the glassware produced by the New England Glass Company were displayed at a local Toledo store, and described as "the finest ever exhibited in this city. Some of the pieces cost as high as $30, and shine with the brilliancy of diamonds."[27]

The glass factory building was designed by architect Charles P. Hamilton, who had previously designed glass factories in Pennsylvania and West Virginia. Hamilton caused some ill feelings within the Toledo business community by criticizing the company's fuel prices as not being as promised to Libbey. Libbey was forced to respond in the *Toledo Daily Blade*, stating he was confident that the gas prices and supplies would be cheap and plentiful.[28] He also indicated that the glass furnaces in the factory were constructed to use gas, oil, or coal in order to control costs. The main building of the Libbey factory was a three-story structure "practically fire proof" and built of brick, iron, stone, and timber.[29] In addition to the glass furnace, the factory included a lehr room (where the glass cooled), mold room, packing room, and storage shed. Over 100 men were employed to build the factory, which was to be the first of several glass factories to be built in Glassboro. Additional factories were planned by the Hurrle Window Glass Works and the McLean Glass Works. One problem with the location, however, was the lack of what architect Hamilton described as "pleasant homes" near the factory for the workers. But despite this concern, the *Daily Blade* proclaimed "Great is Glassboro!"[30]

Glass Comes to Toledo

The arrival of Edward Drummond Libbey and his workers was a time of great celebration in Toledo. "Right Royal Reception Given to the King of Kings Upon Their Entrance to Toledo," the *Daily Blade* proclaimed.[31] The train bringing the workers and company executives arrived later in the day than expected on August 17, 1888. Despite the delay, it was met by the Grand Army of the Republic band and hundreds of Toledo citizens. "Bells were ringing, whistles were blowing, and everyone indulged in a general jubilee," the newspaper noted.[32] The mayor and others officials met the train to welcome the workers.

A parade of carriages ferried Libbey and his superintendents from the depot to the new factory. The parade was led by several foreman of the factory, and the 115 workers carried signs and banners proclaiming their arrival. "From Boston to Boss-Town, We Have Come to Stay."[33] The workers walked the four miles to the factory after the long 36-hour train ride. They were rewarded with tables of food and kegs of beer. One of the oldest employees to make the trek to Toledo noted that the city seemed to have the best beer in the world.[34]

In his address to the crowd, Mayor Kent Hamilton felt it necessary to address the rumors concerning the unhealthy environment of Toledo, which was still trying to cast off its reputation as a disease-ridden cesspool. "We also have one of the most healthy cities on the continent," Hamilton said. "Statistics extending over a period of years show that while the average death rate is fully 20 per thousand per annum, in Toledo the death rate is only 12 per thousand. You have doubtless heard evil reports of the health of our city, but these figures will bear investigation. Because of this it is evident that we have a good soil, salubrious climate, and a city to which you can bring your families," the mayor added.[35]

Edward Drummond Libbey was clearly moved by the outpouring of support for his company from Toledoans. When it was his turn to speak, Libbey said, "Had I all eloquence, I could scarcely do justice to the occasion. . . . I wish I could find words to express our thanks to the people of Toledo tonight for their royal welcome. We want to say upon the very threshold of our arrival that we have come to stay."[36] Libbey presented the mayor with the first piece of glass produced from the furnaces of his new factory.

The factory's furnaces were fired up for full production on August 21, just four days after the workers had arrived in the city. Its initial products

Workers outside the Libbey factory, ca. 1900. (Ward M. Canaday Center for Special Collections.)

included utilitarian wares such as lamp globes, accessories for soda fountains, and battery jars. It also included more artistic products such as lamp shades that gave the workers a chance to show off their fine glass-blowing talents and skill at producing colorful glass. Workers who were interviewed on that first day of production said they were happy with their new home, but expressed concern about the possibility of an end to protective tariffs on glass, which they feared would spell the demise of the domestic industry.[37]

The early years of Libbey's factory were different, however, from the hype that characterized its first few weeks of operation. The furnace did not work as planned. The batch recipes that had produced the brilliant colors that made the New England Glass Company famous could not be matched in Toledo. Some 35 of the men who had come with Libbey from East Cambridge returned home. Libbey had to borrow money from Toledo banks to stay afloat after losing money in 1888 and 1889. By January 1890, the company was $3,000 in debt.[38]

The Libbey factory in the Glassboro district of Toledo, ca. 1888. (Ward M. Canaday Center for Special Collections.)

Libbey was also surprised that the cost of fuel for his factory was more than he had been led to believe. In a public spat with William Maher that played out in the pages of the *Daily Blade*, Maher asserted that Libbey's claim to have paid $5,000 for fuel in 1889 was not true.[39] Maher said the actual cost was $4,200. He also argued that Libbey had not come to Toledo because of the price of gas, and that other towns like Fostoria and Findlay had offered him better gas price deals. Maher said Libbey had come to Toledo because his men refused to move to small towns like these. He also reminded Libbey of the sacrifices that had been made by Toledo to get the company here: "We raised $4000 in hard cash, and bought him a four-acre building site, and in addition gave him real estate in various parts of the city for the nominal value of $15,000. I do not blame Mr. Libbey for trying to get the very lowest rates [for fuel] for the future that he possibly can, but the city covers a block or two more than the glass factory, and the price of fuel ought to be made on a broader basis than by its effects on one factory."[40] Libbey was quoted in the Fostoria paper saying that if the price of natural gas rose any more, he would be compelled to close his factory and move elsewhere.[41]

Fortunately, events interceded to stop Libbey from doing so. In 1890, workers at the Corning Glass Works in Corning, New York, went out on strike seeking union recognition by the AFGWU. As a result, Corning was unable to meet the needs of Edison General Electric to produce the incandescent lightbulbs that were in great demand. Libbey was given the contract for the bulbs. He leased a closed glass factory in Findlay, and hired away some of the Corning workers to come to Findlay to produce the lightbulbs. The contract was profitable enough to pull Libbey out of debt. But while the contract saved the company, it came at a human cost. On July 3, 1891, workers who had come from Corning to work in the Findlay bulb factory were returning to their homes in New York for the holiday weekend when the train they were on wrecked near Ravenna, Ohio. Eighteen workers were killed.[42]

Solon Richardson, one of the executives who had come from Massachusetts with Libbey, managed the lightbulb factory. The manager of production was a glassworker recently recruited from the Hobbs & Brockunier glassworks in Wheeling, West Virginia—Michael Owens. While the lightbulb contract was immediately instrumental in Libbey's survival, it would be Michael Owens who would have a more profound and lasting impact on the company's prosperity.

Michael Owens was born in Mason County, Virginia, in 1859 to Irish immigrant parents who had come to America in the 1840s. His parents moved to Wheeling, West Virginia, and at the age of eight Owens began helping his father in the coal mines. Two years later, he got a job shoveling coal into the furnaces at the glass factory in Wheeling, where he worked a 10-hour day. He worked his way through the ranks, to being a carry-in boy, a carry-out boy, and a mold boy. As he later described his first job of "firing the glory hole," "That was my first job when I was ten. I worked five hours in the morning; and when I came up out of the pit I was black as that ink there. I went home, washed clean, ate dinner, and went back for another five hours in the afternoon."[43] He moved into other jobs in the glass factory that were held by boys. "In the factory, I went through all the jobs which boys performed; and I enjoyed every bit of the experience. I wanted to learn everything there was to be learned."[44] He eventually became a skilled glassblower, an unusual move up for someone who started as a "blowers' dog." Owens credited his upward mobility to the absence of a union that would have impeded his move into skilled jobs. At some point in his early career, however, Owens did join the American Flint Glass Workers Union, and was an active member. He served as a delegate to the union's national meeting in Canton, Ohio, in 1888.[45] Some biographers have claimed that Owens was one of the organizers who led the New England Glass Company strikers that year—the strike that prompted Edward Drummond Libbey to move to Toledo.[46]

The year 1888 was also when Libbey came to Wheeling to look for skilled workers for his Toledo factory to replace those who had come from Boston but who had returned to the East Coast. In an interview done two years before his death, Owens said a friend of his was originally hired as a superintendent by Libbey, but decided to stay in Wheeling. Owens said he then wrote to Libbey asking for the position.[47] While Libbey hired an older glassworker for the superintendent job, Owens came to Toledo to work for Libbey as a glassblower. "From the very first, I knew that the superintendent would not be a success. He was too old a man to take up new methods. He was incapable of handling men who did not know him and who did know far more about the new methods than he knew," Owens recalled. "They were drinking too much, they were slacking their work, and they had lost respect for the man over them. In November, Mr. Libbey let this man go and put me in as superintendent."[48] After taking over the manager job, Owens fired most of the men. "This wasn't a question of men doing badly in one position and perhaps doing well in another. A glass blower was a glass blower; that's all there was to it. If they wouldn't do their work well, I meant to have other men who would. So I got them," Owens stated.[49]

A World's Fair Produces a National Brand

The New England Glass Company had displayed its wares at the Philadelphia Centennial Exposition in 1876 to great acclaim. So when plans were announced for another world's fair to be held in the United States, it was logical to Libbey that his company—which officially changed its name to the Libbey Glass Company in 1892—would want to participate. Edward Drummond Libbey envisioned constructing a working glass furnace at

The Libbey pavilion at the World's Columbian Exposition in Chicago, 1893. (Ward M. Canaday Center for Special Collections.)

the fair similar to the one the Gillinder Glass Works had operated at the 1876 fair.

Libbey approached the company's board of directors to seek funding to build a pavilion at the Chicago fair. The directors were cool to the idea, and in the end, Libbey had to raise the money privately, with the funds secured from investors by Libbey's signature. A separate corporation called the Libbey Glass Company of Illinois was organized and capitalized at $75,000. In the end, the fair's pavilion costs rose to $250,000, and it was described as "the largest glass factory ever built for demonstration purposes."[50]

The pavilion was built on the Midway Plaisance of the fairgrounds. Libbey was given the exclusive concession at the fair to produce and sell American glassware. The display included a working glass furnace managed by Michael Owens that was fueled by oil piped from Ohio. The furnace was in the middle of the pavilion building, glassblowers worked at stations spaced around the furnace, and visitors could watch the workers produce

Michael J. Owens, pictured with glassworkers, ca. 1910. Owens is the man in the middle in the long coat. (Ward M. Canaday Center for Special Collections. Used by permission of Owens-Illinois.)

their wares. The pavilion also included a glass-cutting room where visitors could watch artisans produce the brilliant cut glass that Libbey's company excelled at making.

In addition to these more common products, the pavilion also included something exceptional—a room where glass was spun into fibers. Glass fiber production was still in its infancy, and this demonstration was a major draw. The fibers were produced by heating a glass cane, and then drawing the glass onto a spinning wheel. As described in one of the books written about the fair, "There was a big wheel with a broad, thin metal rim kept cool and moist. The workman sat at one side holding a glass rod before a blowpipe and moving it round and round and slowly forward so as to keep it melted fast enough to feed the single long thread onto the rapidly revolving wheel."[51]

The process for making glass fibers that was demonstrated at the Chicago fair was invented by Herman Hammesfahr, a glassblower from Germany. Hammesfahr came to work at the Libbey plant in Toledo just as Edward

Drummond Libbey was in search of a spectacle that would draw visitors to the Columbian Exposition pavilion. After producing the glass fibers, Hammesfahr taught two women how to weave them into cloth on a standard loom. At the fair, the glass threads were cut into standards lengths once cooled and woven into small pieces such as neckties, doll bodies, and bookmarks that were sold as souvenirs.

The most impressive use of the glass fibers that Libbey's workers produced was displayed in the pavilion's Crystal Art Room—a glass dress. The dress was made for actress Georgia Cayvan, who had visited the exhibition, viewed the examples of lamp shades, ottomans, and tapestries, and requested that the weavers make her a dress of the fibers. The dress was sewn by Herman Hammesfahr's wife. Cayvan was also granted the exclusive rights by Libbey to wear a glass dress on stage. When completed, the Cayvan dress was put on display, and it attracted the attention of Princess Eulalia of Spain, who requested a similar dress for herself. The princess's dress was sewn by a designer in New York, and cost $2,500 to make. It was given as a gift to the princess, who thanked Libbey Glass by recognizing the company as the official "Glass Cutters to Her Royal Highness Infanta Dona Eulalia of Spain."[52]

When the fair first opened, entrance to the Libbey Pavilion was free, but few attended. Libbey decided to start charging visitors 10 cents, and ironically the crowds grew.[53] The crowds became too large for the building, so Libbey raised the fee to 25 cents. The admission fee could be applied by fairgoers to the purchase of Libbey souvenirs. In addition to the trinkets made of spun glass, the souvenirs included objects such as small glass hatchets and commemorative paperweights. All of the items included the trademark of Libbey Glass.

The World's Columbian Exposition captured the imagination of the country. In its six-month run, it attracted nearly half of the country's population—27 million people.[54] The buildings were built to resemble the great buildings of classical Greece and Rome, and most were made of white plaster, like buildings on a movie set. Because of the plaster facades, the fairgrounds became known as the "White City." The landscape of the fair was laid out by Frederick Law Olmsted, the designer of Central Park, and it included buildings designed by Louis Sullivan and sculptures by Augustus Saint-Gaudens. The fair's buildings were lit up at night with electricity.

The fair also ushered in the American consumerism that would dominate the twentieth century. Many of the companies that became recognized national brands were promoted at the fair, including Libbey Glass. Libbey commissioned noted author Kate Field to write a book about the

Libbey Glass display, which she titled *The Drama of Glass*.[55] The book not only celebrated the Chicago fair, but also Libbey exhibitions at the San Francisco Midwinter Fair and the Atlanta Exposition. It concluded that "the Libbey Glass Company thus stands today to represent the best the United States produces in cut glass, and the best the United States produces is the world's best."[56]

Libbey's considerable gamble of creating the glass pavilion at the Chicago fair paid off—extremely well. After returning the money he had borrowed from investors, the costs of the concession rights to the fair promoters, and deducting the costs of construction and operation, the pavilion turned a profit of $100,000.[57] Each cheap souvenir purchased at the pavilion returned home with fairgoers who had come from all over the world. Each included the Libbey trademark. The World's Columbian Exposition not only made money for the Libbey company, but also created a national brand that allowed the company to move into its next phase of development.

Brilliant Men, Brilliant Glass

When Michael Owens returned from managing the glass factory at the World's Columbian Exposition, he turned his attention toward ideas to automate glass production. Edward Drummond Libbey still owned a controlling interest in the company that bore his name, but he left daily operations to Solon G. Richardson, Jefferson Robinson, and William Donovan in 1898.

Libbey's interest moved elsewhere. In June 1890, he had married Florence Scott, the daughter of Maurice Scott and granddaughter of "Future Great City of the World" author Jesup Scott. Maurice Scott was a wealthy Toledo businessman and real estate investor, and was one of the most prominent men in the city. His daughter had been educated at private boarding schools, and had traveled extensively. Her marriage to Libbey brought Libbey prestige and entry into Toledo's high society, along with considerable wealth. Florence Libbey also brought with her a love of art, which she passed on to her husband. When she married Libbey in her late twenties, she had already begun her own art collection, and had grown up in a home surrounded by fine art.[58]

In the years following the Chicago fair, Libbey Glass expanded rapidly. Between 1890 and 1920, the assets of the company increased 26-fold.[59] The company became known for its "brilliant" cut glass, which it continued to showcase on a national stage. This was the time period when cut glass reached its artistic peak. Lathes run by electric power could make deeper, sharper cuts in the glass, producing the faceted shine that brought the glass its "brilliant" description. It fit the aesthetic sense of the high Victorian period.

The cutting of the glass began with a heavy, polished blank produced by highly skilled glassblowers.[60] A designer marked the blank with chalk to indicate the desired design, and the cutter held the blank against an iron miter wheel revolving on a spindle. The glass was actually cut not by the wheel but by a stream of fine sand and water dripping on the wheel. After cutting, the glass had to be highly polished to remove the glass dust.

In 1898, Libbey Glass produced one of the largest pieces of glass it had made to date—a punchbowl for the newly elected U.S. president from Ohio, William McKinley. It took two skilled glass cutters three weeks to produce the bowl, and it weighed 50 pounds. It was decorated with a patriotic pattern of shields, stars, and stripes. Mrs. McKinley described it as the prettiest piece of cut glass ever produced.[61] The president was particularly pleased that it had been made by an Ohio company.

In 1904, the company established a display at the St. Louis World's Fair in the Golden Pavilion, sponsored by Mermod and Jaccard Jewlers. While the display was considerably smaller than that at the Chicago fair, the products were perhaps more impressive. One piece from Libbey was a cut glass table that was 32 inches high with a top 28 inches in diameter. It was made in three pieces that fit together perfectly without any metal fittings. Another display at the fair was a reproduction of John J. Boyle's allegorical frieze that had been on the portal of the transportation center at the World's Columbian Exposition, done in etched glass.

But perhaps the most impressive item was a massive punch bowl that was 25 inches across—twice the size of the largest punch bowl that had been made to date. It was magnificently cut, and was featured on the cover of *Scientific American* in April 1904.[62] It was called the largest piece of cut glass in the world. Before it was cut, the blank weighed 143 pounds. After cutting, it was a third lighter. The cutting was done by John R. Denman, one of Libbey's best glass cutters, and the piece was valued at $2,500. Another punch bowl of note was one that while smaller in size, was no less impressive. It was elaborately etched with a hunting scene that appeared nearly three dimensional. An article in the *Blade* about the works prepared for the St. Louis fair noted that "the greatest care is being taken in the packing for shipment of this exhibit, and Libbey company officials will no doubt heave a sigh of relief when they learn that the precious cargo has safely reached its destination."[63]

Libbey Glass marketed its fine cut glass as the perfect special gift for occasions like weddings. In a booklet the company produced in 1905 titled "The Gentle Art of Giving," it extolled the message that was conveyed by giving a gift of Libbey cut glass. "There is no gift that in itself combines as much of grace and beauty, that appeals to the feminine heart as pleasantly as Cut Glass. Its chaste beauty and sparkling brilliancy possess a subtle fascination that is irresistible."[64]

In addition to beautiful cut glass, the Libbey company had begun experimenting with new processes for producing glass that would influence the entire industry. Michael Owens was working on a machine to automatically make bottles. An adaptation of that machine, called the Westlake machine, produced lightbulbs and tumblers for Libbey using paste molds that left no seams in the glassware. That machine was spun off as its own company in 1907—the Westlake Machine Company—and it achieved commercial success by 1916. In 1915, the Libbey company promoted "Lightware," a thinner, cheaper glass that stood in contrast to its heavy, leaded cut glass. In 1917, the company attempted to reintroduce its Amberina line that had been so successful in the 1880s, but it was unsuccessful.

The popularity of cut glass began to decline around the time of World War I. Some of the chemicals required for the leaded glass were not available during the war, and labor costs for skilled cutters were high. The quality of cut glass produced by Libbey began to decline, producing a drop in sales. Also contributing to the downturn was the changing aesthetic tastes of the country. Fancy, ornate Victorian furnishings including brilliant cut glass were replaced in the most fashionable homes by those with simpler, cleaner lines.

But by this time, Edward Drummond Libbey's personal wealth was considerable. He had arrived in Toledo in 1888 with only $100,000 in total assets, including the equipment that ran his factory. By 1920, the net worth of all of the glass companies in which he held a controlling interest was over $40 million.[65] In 1920 alone, Libbey Glass made $2.5 million in profits.[66] Michael Owens, who had done so much to ensure Libbey's success, was not nearly as wealthy. He was also not as socially prominent. But despite their different backgrounds, education, and personalities, the lives of Libbey and Owens were entwined. As John Biggers, who knew both men, would later note, "These two men seldom or never crossed paths socially, but had the highest regard and respect for one another, at least for one another's integrity and judgment."[67]

Mr. Libbey's Museum

Edward Drummond Libbey's interests began to drift away from his vocation to his avocation—art. Influenced by his wife Florence, he began to study and collect art. He also came into social circles of others with art interests, including Dr. Frank Wakeley Gunsaulus, who was a trustee of the Art Institute of Chicago. In Toledo, the Tile Club, a group of men who both created and studied art, urged Libbey to create a museum like those recently founded in other Midwestern industrial cities like Chicago, St. Louis, and Detroit.

The Toledo Museum of Art was incorporated on April 18, 1901. This event, which would have a lasting impact the city's future, drew little attention at the time. A two-paragraph story in the *Daily Blade* listed the incorporators, which included Libbey and six other artists, art aficionados, and businessmen.[68] The article noted the purpose of the museum: "to erect, establish, and maintain a gallery for the exhibition of paintings, sculpture and other works of art and a museum of natural and other curiosities and specimens of art and nature, promotive of knowledge to establish and maintain an academy for advancing, improving and promoting painting, sculpture, drawing, architecture and other fine arts, and furnishing instruction therein."[69] In addition to the founders, another 120 people signed the articles of incorporation and each donated $10 to be members. At the group's first meeting in May of that year, Edward Drummond Libbey was elected president.

The first exhibits were held in two rooms of the Gardner Building in downtown Toledo in December 1901. In 1903, Libbey purchased the home of Theophilus Brown at 13th and Madison for $10,000 as a permanent museum and home for its director.[70] But the museum struggled to attract members, and George W. Stevens was appointed director in October 1903 in hopes of adding new exhibits and visitors.

Stevens had no background in art, but appreciated culture. Stevens said an art museum was important to a city like Toledo: "A great manufacturing centre is a prison house unless it provides something for the leisure hours. . . . Hospitals do much; they make sick men well—Museums of Art do much more; they make well men better."[71]

A unique feature of the Toledo Museum of Art under Stevens's direction was an emphasis on education as well as exhibitions. Classes were offered not only in art appreciation, but also art techniques. This was in keeping with turn-of-the-century ideas about education that stressed that both the mind and the hand must be educated together to create the whole being.

The Libbey exhibit at the 1904 St. Louis World's Fair, showing off the company's brilliant cut glass. (Ward M. Canaday Center for Special Collections.)

Schoolchildren were allowed to visit the museum, and talks were given to the general public on a daily basis. Stevens even took his art education lectures to local factories.[72]

Edward Drummond Libbey and his wife began to travel the world, collecting art for their new museum. The Libbeys traveled to Egypt and throughout Europe. In 1907, Libbey acquired a collection of Egyptian antiquities. That year alone 1,400 works were added to the museum's collection.[73] The building on Madison became too small to either exhibit or store the extensive and growing collection, and Libbey offered to sell the building and donate the $50,000 for a new building. But he requested that the city raise an equal amount of money in just a few weeks to help fund a new museum. The city came together and embraced the challenge, with schoolchildren donating pennies and nickels, and others donating what they could. Toledoans met the deadline, and Libbey and his wife donated the land on Monroe Street that had been the homestead of Florence's father, Maurice, for the new museum. The land was valued at $75,000.[74]

Designed in an impressive neoclassical style, the new Toledo Museum of Art opened to the public on January 17, 1912. Because of the large number of people who wanted to get in, the first day was for trustees and members only. But even with this limitation, the crowd of people seeking entrance was so large that the museum had to close its doors when it reached its capacity of 5,000 people. Dr. Frank Gunsaulus was supposed to give an address on the importance of an art museum to a manufacturing city, but given the size of the crowd, decided against it. Toledo mayor Brand Whitlock paid tribute to Libbey, and gave him a gold key to the city. As the *Daily Blade* reported, "Mr. Libbey was visibly affected when he arose to receive this mark of esteem from the representative of the city government, and of his fellow citizens."[75] Whitlock also gave Libbey a set of books containing written testimonials from 40,000 people thanking Libbey for his gift to the city.

The first exhibit at the museum included none of its permanent collection. Instead, museums from around the world loaned some of their most important works for the event, and there was no room for others. The works on display for the opening were valued at over $3 million.[76]

Because of his connection to the glass industry, glass as museum objects became part of the Toledo Museum of Art's permanent collection beginning in 1913, when Libbey acquired 80 pieces of European glassware. The first piece of Libbey glass added to the museum's collection was the punch bowl with the engraved hunting scene that had been made for the 1904 St. Louis World's Fair.

THREE

The Man and His Machine

> The Owens machine stands alone in a class unapproached by other inventors. It differs radically and fundamentally from all other machines in several important and epoch-making features, because it gathers its glass, forms its blank, transfers the blank from the gathering to the blow mold with a finished lip and rings, blows the bottle, and delivers the finished bottle automatically, without the touch of a human hand, eliminates all skill and labor, and reduces the cost of production practically to the cost of materials used. Not only that, but it puts the same amount of glass into every bottle, makes every bottle of the same exact length, finish, weight, shape and capacity. It wastes no glass, uses no pipes, snaps, finishing tools, glory-holes, gatherer, blower, mold boy, snap boy, or finisher, and still makes better bottles, more of them, at a lower cost, than is possible by any other known process.
>
> —Article about the new Owens Bottle Machine appearing in *The National Glass Budget*, August 29, 1903

The technique used to make glass bottles in 1900 differed little from the technique used in AD 50. The craft required skilled glassblowers, a blowpipe, a glass furnace with a "glory hole," an annealing lehr, and at least four unskilled workers, usually young boys. The process used in a typical shop consisted of two blowers and a finisher. A blower would gather just the right amount of molten glass (heated to around 2,400 degrees) from the glory hole on the end of his blowpipe. A young boy, called the "mold boy," would close a bottle mold around the hot glass. The blower would blow the mold,

Child glassworkers, ca. 1900. (Ward M. Canaday Center for Special Collections. Used by permission of Owens-Illinois.)

and the mold boy would open the mold to reveal the bottle. Since the neck of the bottle had to be shaped, it was turned over to the finisher. Another boy would clean off excess glass from the bottle and the mold, and another would snap off the finished bottle. The fourth boy would carry the hot bottle to the annealing lehr for cooling. A typical seven-person shop could produce about 300 dozen eight-ounce bottles a day working 10-hour shifts.

Because the unskilled jobs were filled by boys, the glass industry—particularly the bottle-blowing industry—was one of the largest employers of child labor in the country. In an article appearing in *Charities* magazine in 1903, Florence Kelley, secretary of the National Consumers' League, called the industry "a boy destroying trade."[1] The boys, called "blowers' dogs," were doubly exploited. First, they had to endure the hot and dangerous work. Often they were forced against their will into the factories. Kelley wrote about boys working at the Illinois Glass Company in Alton, Illinois, some

"Blowers' dogs" helping to produce glass bottles, ca. 1900. The glass industry was the largest employer of child labor in the country prior to the automation of the industry. (Ward M. Canaday Center for Special Collections. Used by permission of Owens-Illinois.)

as young as seven years, who had been gathered up from poorhouses and orphanages by disreputable men. Some boys were the children of disabled glassblowers who, because of the difficult work of blowing glass, could no longer work themselves. The boys' pay depended on how quickly they worked in the intolerably hot factories, and they worked nights as well as days. They were often physically and verbally abused by the glassblowers. In addition to these horrendous working conditions, the boys had little chance to move up to become apprentices and eventually well-paid skilled blowers because union contracts limited the number of apprentice positions.

Kelley also noted that child labor laws intended to protect the children were often not enforced, especially in glass towns.[2] When attempts were made to enforce existing laws in Alton in 1893, the Illinois Glass Company told the local newspaper of the dire consequences awaiting widows who depended on the labor of their sons in the glass factory if the laws were enforced. When the Illinois legislature enacted more stringent laws governing work in the bottle factories in 1903 that prohibited night work, reduced the workday to eight hours, and required child workers to be registered, legislators from Alton actively opposed the changes and the Illinois Glass Company threatened to leave the state.

Michael Owens was all too familiar with the life of a "blowers' dog." It is unknown if Owens was inspired to develop his machine to automate glass bottle production because of his experience as a child in the glass factories. But as a result of his automatic bottle machine, child labor finally came to an end in the glass bottle industry.

The Coming of the Machine

Glass bottle production in America in the late nineteenth century was a handicraft industry. Most firms were small in size, and many were family businesses, with trade secrets passed down from father to son.

Glass bottle blowers were one of the first trades to unionize. In 1846, small groups of organized bottle blowers came together to form the Glass Blowers League. That year, they were recognized by the industry as the bargaining agent for League members, making this the first industry-wide collective bargaining agreement in the United States.[3] The League flourished until the depression of 1856 and the Civil War negatively impacted the industry. The League was succeeded in 1868 by the Independent Druggists' Wares League, which existed mainly in Pennsylvania around Pittsburgh. In 1886, the group joined the Knights of Labor. The organization's most important issues were child labor, immigrant labor, and cheap imported glass, all of which tended to deflate the wages of the skilled bottle blowers. In 1890, the workers reorganized again as the United Green Glass Workers' Association as a unit of the Knights of Labor. In 1891, they withdrew from the Knights, and around 1900 dropped the word "Green" from their name, and the union became known as the Glass Bottle Blowers Association (GBBA) with its headquarters in Pittsburgh. The union fought with the AFGWU over organizing rights in the glass factories before the two came to an agreement in 1913 concerning what type of workers could join which union.

Glass bottle blowers were well paid in comparison to other workers, and the union kept tight controls on production and prices. Early union contracts consisted of detailed lists of types of bottles and what blowers were to be paid for each type and what each type was to be sold for by the factories. Labor costs constituted between 41 and 45 percent of the cost of glassware in the late nineteenth century.[4] Manufacturers had an incentive to automate bottle production not only to reduce labor costs, but also to reduce the power of the labor unions.

The first inventor to develop an automated process for bottle production was Philip Arbogast of Pittsburgh. He filed a patent application in 1881 for a press-and-blow process. In this three-step process, the neck of the bottle was pressed into a mold first using a plunger, then the plunger was removed and the bottle removed from the first mold and inserted into a second mold, where the body was blown. Arbogast was unable to make money from his patent, so he sold it for almost nothing to the D. C. Ripley Company of Pittsburgh in 1885. The Ripley Company also was unable to make money from it because its contracts with the union that set production and wages. The patent was sold again, and was eventually controlled by Charles E. Blue of the Wheeling Mold and Foundry Company. It was used to make wide-mouth jars, but was unsuccessful in making narrow-necked bottles. A successful narrow-neck automated process was developed in England on a machine made by Howard M. Ashley and patented in 1886, but it was not profitable and had only limited use.[5]

Michael Owens began to experiment with new methods of glass production in 1885.[6] That year he invented a bottle mold called a "dummy" that could be opened with a foot treadle that allowed the blower to perform this work rather than a boy. It was first used in large-scale production in 1890 when Owens oversaw the production of lightbulbs for Libbey Glass.

But it was upon his return from running the glass factory at the World's Columbian Exposition that Owens began to seriously work on a machine to fully automate bottle production. In 1894, he received patents on two semiautomatic paste-mold machines to produce lightbulbs, lamp chimneys, and tumblers. In this process, the skilled worker attached a gob of glass to a blow pipe, and compressed air blew it into a mold.

Edward Drummond Libbey thought that the patents Owens developed could be profitable if they were licensed to other companies. On December

16, 1895, Libbey and five other men from Libbey Glass created the Toledo Glass Company to profit from Owens's patents. In exchange for rights to his patents for any invention developed during the next 17 years, Owens received 800 shares of the new company's stock. The company's articles of incorporation stated that it was formed, among other purposes, to construct "implements, machinery and mechanical devices" for manufacturing glassware, and "purchasing, selling and dealing in improvements, inventions, processes, patents and rights for and pertaining to such manufacture."[7] With the formation of the Toledo Glass Company, Owens removed himself from Libbey Glass, where he did not get along with E. D. Libbey's principal partners.

Because most Toledo banks thought Owens's experimentations were a poor risk, Libbey himself was the largest source of venture capital for Owens's inventions. The Toledo Glass Company had difficulty selling its initial stock because few investors believed licensing automated glass-blowing machines would be profitable.

With Libbey's money, Owens's technical expertise, legal counsel from Clarence Brown and Frederick Geddes, and administrative oversight from William Walbridge, the company set about to make money from Owens's work. They sold the rights to the semi-automatic tumbler machine to the Rochester Tumbler Company of Pittsburgh. The chimney lamps were produced at the Toledo Glass Company factory under the name of the American Lamp Chimney Company.

News of the Owens inventions quickly spread throughout the glass industry. An article in the *Glass Budget* about the machine used by the American Lamp Chimney Company noted that "the Owens machine enables the combined manufacturers to make about 2,000 chimneys per turn of five hours at a labor cost of $6.50, as against a hand shop product of less than 200 chimneys per turn at a cost of $4.60, the majority of chimney manufacturers pursuing the hand making process cannot possibly compete in the same market."[8] The American Chimney Lamp Company estimated that it could supply half the number of lamp chimneys needed by consumers in the United States using the Owens semi-automatic machine, and it could gradually increase its output to supply the entire market. The article in the *Glass Budget* also made clear the implications of the machine for the skilled glass bottle blower: "The feeling has become general that they can only hold on to a remnant of their trade which must constantly vanish as the machine equipped combination augments production to an extent sufficient to supply the entire demand at prices which is it impossible for inferiorly equipped factories to meet."[9] The machines were first installed in nonunion factories because it was feared that union workers would sabotage them.

The Automatic Bottle Machine Is Born

Because Owens was not an engineer but knew the technical aspects of making glass, in 1899 he worked with engineer W. Emil Bock to begin fashioning a machine that would fully automate the bottle production process. The first machine was called the "bicycle pump."[10] It worked by sucking up the correct amount of molten glass into something similar to a blowpipe, and then reversed the air to force the glass into the neck mold. The neck was kept open by means of a plunger, and another blast of air forced more molten glass into a second mold to form the body. The most important improvement in this machine over other attempts to make a bottle machine was that it allowed for successfully making narrow-necked bottles. But the machine was crude. While Owens did not receive a patent for this machine until 1904, it was the basis for more refined iterations that became known as the Owens Bottle Machine.

Working with Bock in a factory on Wall Street in Toledo, Owens had to solve several problems to perfect the machine. One was that the machine stopped and started with each dip into the pot of molten glass. This slowed production considerably. The second was that with each dip into the pot, the glass cooled as it came in contact with the mold so that it could not be drawn correctly up into the entire mold. Owens solved these problems by inventing a revolving pot that moved around the machine and maintained the exact temperature required. Then Owens decided that rather than move the pot, the machine itself would rotate continuously.

The next improvement involved installing multiple arms on the machine, each of which could hold a different mold. The first "A" model had six arms mounted on a central column. It could produce 12 bottles per minute. Later, Bock and Owens expanded the machine to 8, 10, and even 15 arms. With these large-scale automation processes, Owens found the furnaces in use at glass factories of the time were too small to accommodate expanded production. He developed an improved continuous tank furnace intended to run large production batches without interruption. Owens also introduced instruments to measure the temperature of the molten glass, something that old-time glass producers relied on their personal experience to tell them. New methods for making molds were also required so that they could produce bottles to exact specifications each time. As Richard LaFrance, who would replace Bock as Owens's chief engineer, noted years later, "While we mention here the 'MACHINE,' it is a fact that the whole process of bottle manufacture, including such units as gas producers, furnaces, mixing plants for materials entering into glass making, annealing ovens for the bottles, and the machine shop equipment for making molds was as crude as can be imagined. Mr. Owens's driving force hammered these crude elements, each one of which at times seemed like an insurmountable wall, into a productive industry. A NEW INDUSTRY HAD IN FACT BEEN BROUGHT INTO EXISTENCE around his dream of making perfect bottles automatically and cheaply, a thing which experts at the time said could not be done!"[11]

The Owens Bottle Machine received its patent in 1903. Future improvements included the automatic loading of the ingredients for the glass pots and a conveyor that would move the bottles into the annealing lehr so they did not have to be carried.

The success of the machine led the Toledo Glass Company to spin off a new company focused on improving and selling licenses to the Owens Bottle Machine. At a special meeting of shareholders held on September 16, 1903, Libbey announced that "the Owens bottle machine had been so far perfected as to manufacture bottles, and that after years of efforts and experiments, many failures and discouragements, and the making of enormous expenditures (all without calling upon stockholders generally for contributions or assistance) the machine promises to be a commercial success."[12] One month later, the Toledo Glass Company assigned all of the rights it owned to the Owens inventions to the new company, named the Owens Bottle Machine Company.

A Company for the Machine

Early accounts of the Owens Bottle Machine that circulated throughout the industry marveled at its accomplishments. "At present we desire to say that the Owens bottle blowing machine is a mechanical marvel which is certain to rapidly revolutionize the entire bottle, jar and prescription industry, and most effectively and economically solves the problem of automatic vacuum glass gathering, blowing, and finishing."[13] And another description: "No one can view and study the Owens machine and witness its performance without realizing that it is the creation of a resourceful and practical glassmaker; that some master mechanic has had a hand in its construction who had a wide knowledge of mechanical movements, the ability to compactly assemble means to secure results, a mathematical appreciation of proportion, and a masterly grasp of diversified mechanical detail, and last, but not least, that there must have stood somewhere in the shadow, unseen but not unfelt, some firm-nerved man with faith and a bank account, whose means

Michael Owens in Manchester, England, at the Owens Bottle Machine Company's model plant, 1906. (Ward M. Canaday Center for Special Collections. Used by permission of Owens-Illinois.)

and liberality enabled that combination of inventive genius and constructive ability to slowly and patiently evolve, despite many mistakes and reverses, that modern mechanical marvel known as the Owens automatic glass working machine."[14]

The business model for the Owens Bottle Machine Company focused not on producing bottles, but on producing and licensing the machines that made them. Because of this, the bottle production facilities in Toledo were small. The machinery itself was made by the Kent Machine Company of Toledo. Libbey believed that selling and controlling the licensing of the machines was where the most profit could be realized. The company began commercial licensing of the machines in 1904.

One of the first expansions of markets attempted by the Owens Bottle Machine Company was overseas. In 1905, the Owens European Bottle Machine Company was established to sell machines and licensing agreements in a market that was much less automated than the U.S. market. To

The Owens Bottle Machine had over 10,000 separate parts. (Ward M. Canaday Center for Special Collections. Used by permission of Owens-Illinois.)

show off its machines, a demonstration bottle plant was built on a 20-acre site in Trafford Park in Manchester, England, in 1906. The factory had two furnaces and two six-arm Owens machines, and Owens and Bock themselves oversaw installation, with William Boshart managing the factory. They were prepared to expand the bottle production at the plant if the European manufacturers could not be convinced to acquire licensing for the machines. If that should have happened, the factory would have turned into a bottle production facility rather than a demonstration plant for the machines.

Operation at the British plant began in April 1907. Many European manufacturers were skeptical that bottles could be produced automatically until they witnessed the machine in production. Recognizing the threat posed by the Owens machine, a European syndicate of glass producers quickly purchased the rights. Owens machines were installed in Germany in 1907, Holland and Austria in 1908, Sweden and France in 1910, Denmark and Italy in 1912, Norway in 1913, and Hungary in 1914.[15]

In the United States, the business model of the company was complex. The company issued its first license to the Ohio Bottle Company (later called the American Bottle Company), a conglomeration of several smaller producers that included factories in Massillon and Wooster and had a large plant in Newark, Ohio. The company was sold the exclusive rights to produce beer, beverage, water, grape juice, liquor, and wine bottles using the Owens machine. The first royalty payment paid to the Owens Bottle Machine Company was by the Ohio Bottle Company, and it was for $1,337 in August 1905. Another early license went to the Northwestern Ohio Bottle Company, which was located at the corner of Detroit, Phillips, and Sylvania avenues in west Toledo. The Owens Bottle Machine Company continued to expand its machine production to allow for producing other kinds of bottles, and as each was perfected, licenses were sold to other bottle producers. One of the earliest bottle producers to purchase rights to the Owens machine was the Illinois Glass Company in Alton, Illinois. Others were sold to the Hazel-Atlas Glass Company in Wheeling, West Virginia; the Ball Brothers of Muncie, Indiana; and the Charles Boldt Glass Company of Cincinnati. Canadian rights were sold to the Dominion Glass Company Ltd. of Montreal.[16]

Emil Bock left the Owens Bottle Machine Company in 1910 to go into business for himself. Richard LaFrance, who had been Bock's drafting assistant, was put in charge of further developments of the machine. Bock felt that there remained little to improve about the machine, and that LaFrance would simply oversee general engineering issues. But Owens was constantly tinkering with his machine.

The "A" machine was followed by the "AC" and the "AD" machines. More heads were added, with the "AD" machine having 10 of them. This added considerable weight, and as it lowered and dipped into the pot at ever greater speeds, the machines vibrated excessively. This was solved in 1912 when the "AN" machine was introduced with individual suction molds that dipped one at a time into the pot rather than the entire machine. The machine allowed for installing 10 different molds, and could make between 50 and 60 bottles a minute. This was quickly replaced by the "AR" machine and the "AQ" machine, the latter with 15 arms.[17]

Improvements to the machine were evident in the capacity to produce bottles of each successive model. The "A" machine—the first one sold commercially—had six arms and could produce 10–20 bottles per minute, depending on size. The "AD" machine had 10 arms and could produce 24–26 bottles per minute. The "AV" machine, introduced in 1917, had 15 arms and could produce 30–90 bottles per minute. A machine to produce carboys, or very large bottles that could be up to 14 gallons in size, was introduced in 1915 and could produce between six and seven such containers per minute. Because of the limited market for large containers, only one carboy machine was ever built, and it was installed at the Illinois Glass Company.[18]

Man versus the Machine

The journeyman glassblower in 1900 was a skilled craftsman. He turned bottle production into a science. Because they were paid by the piece, glass bottle blowers worked 10-hour days. While the days were long, the pay was good—a glassblower could make up to $45 a week.[19] The first change in the status of the skilled worker was seen in 1898 with the introduction of the semi-automatic machine. The union had to accept a 45 percent cut in the piece rate that year, and it hoped that some of the displaced skilled workers would get jobs running the machines.[20] With the installation of the completely automatic bottle machine, it was not unusual for the skilled glassblower to see his pay to shrink to 25 cents an hour for a ten-hour day.

Michael Owens had been a member of the AFGWU, and served as a delegate to the union's national conference in 1888. Despite this, the stock prospectus for the Owens Automatic Bottle Machine issued in 1903 touted not only the labor costs savings that the machine could produce, but also that it could cripple the power of the glass unions.[21] The *National Glass Budget* stated: "While the absolute elimination of all skilled high price labor is the main feature of the Owens machine, there are additional savings which will readily suggest themselves to all practical manufacturers." These included the ability to run the glass furnaces 24 hours a day, the end of the two-month summer shutdown, and that the machine "resorts to no strike for increased wages or less work."[22]

In a publication entitled "The Owens Automatic Bottle Machine: Its Relation to the Bottle Industry of the United States," published by the Owens Bottle Machine Company around 1904, the stark comparison between skilled labor and the machine was made clear. It noted that over 60 percent of the cost of making bottles went to skilled labor. "In the use of the OWENS AUTOMATIC BOTTLE MACHINE, it will run continuously twenty-four hours, knows no holidays, is not affected by warm weather, does not depend on glassworkers or boys, and it can certainly never strike."[23] The publication noted that three machines would require just three workers during a 24-hour shift as opposed to between 108 and 167 skilled workers and boys to make bottles by hand. Labor costs could be reduced from $396 per 18-hour day to just $10 per day for a 24-hour shift. And since the fully automatic machine required no skill to operate, the workers could be paid much less.

The Glass Bottle Blowers Association took notice of the machine, and tried to find a way to fight it. When one of the fully automatic machines was installed at the Ohio Bottle Company, workers sabotaged it. At its 1906 convention, the union suggested a three-part plan to combat the machine and save the union. The first point was to expand the union throughout the glass bottle industry, ensuring that bottle blowers were represented by the GBBA and not the AFGWU. The second was to reduce working hours in order to keep employed as many workers as possible. As the convention noted, "This Association would likely prefer three shifts to seeing good workmen and citizens made idle through a heartless industrial system."[24] The third point was to bring all nonunion bottle factories into the GBBA.

At the association's 1908 convention, the union actively advocated for the eight-hour day because it would reduce work hours, put more men to work, still allow for 24-hour continuous production, and help keep the workers and the producers competitive.[25] The association also reported an increase in the number of automatic machines that were currently being used in the industry, which was up to 36. Of these, 27 were used at the American Bottle Company factory in Newark, Ohio. By 1910, that number was up to 60, with 36 more ready to be installed. The association noted that if production of all automatic machines then in use was kept up for a year, the number of bottles produced by the machines would top one million. In its publication *American Flint* in July 1910, the AFGWU suggested that the union's members should consider purchasing the automatic bottle machine, thus controlling its use and allowing it only to be used in competition with nonunion factories.[26] But the members did recognize that the cost was prohibitive, and even if this were possible, new machines would likely be developed.

T. W. Rowe, president of the AFGWU in 1910, had worked with Michael Owens at the Libbey plant in Toledo. He seemed almost apologetic that the bottle machine had been invented by a former AFGWU member. But Rowe was clearly impressed by the new invention. He also warned his members against assuming the machine would not last. "A dream of that kind may be consoling, yet I advise that some one wind the alarm clock and set it for an immediate ring," Rowe stated.[27]

Despite the impact of the machine on union members, the union continued to express deep respect for Owens. The 1910 convention of the AFGWU was held in Toledo, and Owens and his wife arranged for an automobile tour of the city for those attending. Owens was graciously thanked for his hospitality in the convention proceedings.

But neither the AFGWU nor the GBBA could deny the swift, lasting, and profound impact of the Owens machine on their members. The number of skilled bottle blowers declined from 8,000 in 1905 to 4,000 in 1917.[28] Those still employed in hand-making bottles produced unique, limited quantity bottles like those for perfume where consumers were more likely to see the bottle as a work of art than as simple receptacle. Summer stops, when glass factories shut down for two months during the hottest time of the summer and which had been a part of glass blowing for centuries, ended industry-wide by 1917.

The unions were required to make four major policy changes as a result of the mass introduction of the bottle machine.[29] The eight-hour workday was adopted. The number of apprentices used in the industry declined dramatically, and by 1909, manufacturers agreed not to accept any more apprentices into the factories. The unions saw the semi-automatic machines as a lesser evil than the fully automatic Owens machine, so they encouraged their use. Lastly, the unions regularly agreed to cuts in their piece-rate wages in order to bring skilled wages into line with the costs of machine production. The GBBA agreed to wage cuts in 1909, 1912, and 1914. It soon became obvious, however, that this was not a good strategy. Even Michael Owens urged the unions not to accept pay cuts in 1912. Owens saw that as the workers cut their pay, the price for bottles made on his machines also dropped, and the company was unable to make as much through its licensing agreements with manufacturers.

Another labor change was the introduction of women into the glass bottle workforce. With large numbers of bottles being produced, there was a need for bottle sorters and packers. Women were employed in this work because "they possess a nicety of selection and a keenness of eye far superior to that of the average male."[30]

But by far the biggest impact brought about by the introduction of the Owens Bottle Machine was on child labor in the industry. While skilled bottle workers were dependent on children's labor—and exploited them—the automatic machine allowed the union to finally come out in strong opposition to the employment of children in the glass factories. Beyond the altruistic reasons to support ending child labor, it was also true that children in the glass factories depressed wages for all. Now that skilled workers were fighting to remain employed at even dramatically declining wages, children were a further threat to the union. In an article in the *American Flint* in 1911, the magazine urged the union to demand the abolishment of child labor. "According to the reports of labor inspectors the state of health is distressing; 'Pale faces, dim and inflamed eyes, inflated bodies, swollen cheeks, distinguished the children so early estranged from their families, which have passed their youth in grief and misery, from those of other classes not compelled to work in the factories,'" the article stated.[31] When AFGWU president William Clarke addressed the Toledo Rotary in 1925, he talked about how he began working in the glass factories in Toledo as a child. "I went to work when I was 9 ½ years. I am here to say that we suffered cruelties. And if any man were to impose those cruelties on a child of mine I would probably have to settle for a crime," Clarke told the Rotary audience.[32]

In a letter written in 1913, the National Child Labor Committee thanked Michael Owens for helping to eliminate child labor. Some factories were still using blowers' dogs to make handmade bottles and bottles made on semiautomatic machines, but by 1920 almost all such positions were automated and eliminated. In 1880, children accounted for 23.4 percent of the workforce in the glass industry overall. By 1919, that had declined to 1.8 percent. The U.S. Department of Labor Statistics, in a report issued in 1927, noted that "child labor in the glass industry has not become almost a thing of the past, and credit for this is due in no small measure to Michael J. Owens."[33]

The Machine or the Bottle

The Owens Bottle Machine Company continued to expand in its early years. In addition to continuously upgrading the machine to make it more efficient and marketing each new development to the factories licensing the machines, the company also expanded its own production of bottles. Prior to 1908, the company did little actual bottle production, and instead focused almost exclusively on building and licensing machines. In 1908, the company purchased all of the stock of the Northwestern Ohio Bottle Machine Company in west Toledo to produce bottle types that were not currently under license to other companies. In 1909, the Owens Bottle Machine Company began building a large bottle production plant in Fairmont, West Virginia, and considered expanding bottle production at its other plants. "From your Company's experience, your Directors are satisfied that the manufacture of lines of bottles, not already licensed, can be conducted at a substantial profit over the royalties which they could reasonably expect, and, therefore, if satisfactory licenses cannot be speedily made, that your Company's revenues will be increased by itself engaging in the manufacturing of bottles," the annual report for that year noted.[34] By 1910, the company reported that its Fairmont plant was in operation and expected to produce 100,000 gross of bottles annually.

In 1916, the change in the company's business model was complete. The Owens Bottle Machine Company purchased outright the American Bottle Company in Newark, Ohio. In a span of eight years, the company had moved from almost exclusively making and licensing bottle machines to producing bottles in ever larger quantities. The company was still making money off its licensing and royalty agreements, with over 100 Owens machines in 15 factories including the Ball Brothers Glass Manufacturing Company, the H.J. Heinz Company, and the Illinois Glass Company. Those agreements produced income of just over $600,000. But the purchase of the American Bottle Company moved the Owens company from being predominantly a producer of machines to a producer of bottles. By 1918, the company was making more profits from bottle production than from selling machine licenses.

The transition was made clear in 1919 when shareholders approved a change in the company's name from the "Owens Bottle Machine Company" to the "Owens Bottle Company." It developed its own trademark for its bottles: an "O" within a square, which was incorporated onto all its products. By that time, the company had bottle factories in production in Toledo; west Toledo; Fairmont, West Virginia; Clarksburg, West Virginia; Charleston, West Virginia; Greenfield, Indiana; Glassboro, New Jersey; Newark, Ohio; Streator, Illinois; Evansville, Indiana; and Loogootee, Indiana. In 1919 the company also acquired outright the Charles Boldt Glass Company in Cincinnati along with 71 patents it owned for its machines. The success of the company had made its stock attractive to investors outside of the small group of Toledo glass men who had originally invested in the Owens Bottle Machine Company.

The Owens bottle and the machine that made it became a symbol of American ingenuity. At the fourth annual convention of salesmen for the Owens Bottle Company in 1923, an unknown member of the sales staff crafted a poem that summed up the significance of Michael Owens's invention to history:

Towering I stand
And gaze around.
The sky above
Below the ground.
None of my kind
In vision's range,
Except below
Where all seem strange.
Not of my kind
They seem to be,
But rather
Some strange fantasy.
I read by fate
To stand alone,
When all the rest
Have gone, have gone.
I will alone
In grandeur stand,
To serve mankind
Throughout the land.
I gave through mist
Of years to come,
And see myself
Excelled by none.
Serving your children
As I serve you,
And their children's children
The ages through.
Strong as the mountain peak
I stand,
Alone, alone
In solitude grand.
The years will come
With naught to throttle,
I stand alone
An Owens Bottle.[35]

The Man and His Machine

Owens continued to be directly involved in improving the machine, working closely with Richard LaFrance, the company's chief engineer. Together, the two worked on expanding production capacity, eventually developing a machine that could hold 20 molds at once. But Owens was still not happy, and asked LaFrance to work on a machine that would hold two molds side by side at right angles on each head so that each mold was the same distance from the center of the machine for even glass temperature. This machine, called the "CA" machine, was extremely complex and expensive to produce. It also required an extra large furnace. By this time in its evolution, the Owens Bottle Machine weighed 100,000 pounds and was composed of 10,000 separate parts.

LaFrance did as requested, and in 1922 the machine was completed but was largely impossible to operate.[36] Despite the difficulties with the CA machine, Owens was determined to put it into production at the Charleston, West Virginia, plant. At a meeting with the company's board of directors on December 27, 1923, Owens strongly stated his case for the machine. The directors were not moved. It was the last meeting of the Owens Bottle Company directors that Owens would ever attend.

All models of the Owens's bottle machine were suction machines because the glass was sucked up into the molds. While this was the industry standard, the Graham Glass Company in Evansville began experimenting in a "flow" machine, where glass was fed into the mold through a flow rather than suction process. Owens had never considered that a method other than suction could be successful. But because the Graham machine showed promise, the Owens Bottle Company purchased the Graham Glass Company and its patents in 1916.

The technical problems with the "CA" machine and the unwillingness of Owens to experiment with anything but a suction machine pointed to the difficulties the Owens Bottle Company began to experience at this time. Most were the result of one man—Michael Owens. By many accounts, Owens was a difficult person. In his lifetime, he moved from being a blowers' dog to a factory-floor glassblower to being a major shareholder in a company bearing his name. Some said he was more comfortable with glassworkers than with the executives who ran the company. The glassworkers, in turn, always treated him with respect when he walked through the factory. LaFrance described Owens as "a generous and appreciative boss, the biggest man I ever knew, a leader among men."[37]

As the Owens Machine Company began to shift its focus from developing and licensing Owens's inventions, and toward bottle production, Michael Owens's relationships with its directors began to strain.[38] At the request of stockholders, the number of company directors expanded, but the board soon became unwieldy. A smaller executive committee of Owens, Libbey, Clarence Brown, William Walbridge, William Boshart, and John Biggers was established in December 1916 to overcome the problem of the larger board.

But Owens, whose background was in glass technology development and not in business management, was not happy. He was impatient, and it was up to Libbey to try to direct this impatience. In 1916, Owens recommended dissolving the Toledo Glass Company—the company Libbey had formed to exploit Owens's inventions and technical innovations. In 1917, Owens resigned as president and director of the company. When Libbey resigned as president of the Owens Bottle Company that year, he named Clarence Brown as president, not Owens, which Owens saw as a slight. By 1918, Libbey had turned his attention away from the bottle company and was spending more time running the new sheet glass company that he and Owens had started. Libbey was also spending more time at his other home in Ojai, California.

By 1919, the Owens Bottle Company was so out of control that Libbey had to issue a letter "to all officials, heads of departments, superintendents, office and factory employees" insisting on company loyalty.[39] The letter threatened that anyone divulging information about the company to anyone outside of it would be fired. It also demanded an end to stock speculation by Owens Bottle Company employees, something Owens had been complaining about for some time.

Although Owens had fewer official roles in the company by this time, he still had considerable influence. In 1921, Owens demanded that the Owens Bottle Company pay him and his estate one cent for each gross of bottles made on any of his machines for the next 17 years, up to $100,000 a year. In making his demands in a letter to Edward Drummond Libbey, Owens stated, "Other inventors are handsomely compensated both by ourselves and by our competitors for inventions that are trivial in comparison [to the bottle machine]. On the other hand, I have not been suitably compensated for inventions in proportion to the profits that have come into the treasury of The Owens Bottle Company, by way of royalties, cash sales or rights or our own use of them."[40] Owens stated he had not discussed this with other company executives, and that he considered it a personal favor if Libbey would consider the demand quickly. Owens said he needed this change because he was reaching the age where he would no longer be active in the company, but he still required sufficient income.

Another factor in Owens's demand may have been his declining health. In 1917, he began to be treated by Dr. Nathan Brown, the foremost Toledo cardiologist of the time, for a weakened heart.

One of the lingering mysteries about Michael Owens was how much of a role he actually played in inventing the Owens Bottle Machine. John Biggers, who started working as a glass company executive for Owens and who would eventually become president of the Libbey-Owens-Ford Company, said of Owens: "He explained [his ideas] verbally to a draftsman, and the draftsman had to reduce them to paper. . . . At least one of his draftsmen, who'd taken visual concepts and reduced them to mechanical form, thought of himself as the inventor rather than Mr. Owens, and that caused certain problems."[41] Biggers also found it odd that Owens was not mechanical in any way, and never even drove a gasoline automobile, presumably because he could not understand how it worked.

But others have said that it was Owens's grasp of glassmaking that made him the brilliant inventor of the machine. Richard LaFrance, who found Owens to be a demanding boss and stated that some called him the "dictator," made it clear in his memoirs that it was Owens who invented the machine, not Emil Bock, his engineer at the time of the first bottle machine. "I worked under both Bock and Owens in this trying period. I claim that neither would have succeeded alone. They made it, in spite of repeated discouragements, as a team and with the faith E.D. Libbey had in Owens. Many times they might have given up."[42] LaFrance said that while Owens was not a designer, he could direct the engineers. "He did not invent mechanisms. He decided on methods." LaFrance added, "It seems to me that without Owens' initiative, guidance and support, Bock could not have succeeded. It also seems that without Bock's genius, dogged perseverance and patience in trying times the effort would have ended in failure. Let us give them all the credit and not try to determine the percentage due each."[43] In an interview in 1922, just the year before his death, Owens clarified that his inventions were the result of his own work, the work of the engineers who solved the mechanical problems, and the Libbey's financial backing.

In his book *Revolution in Glassmaking*, published in 1948, historian Warren C. Scoville concluded that Owens was a success because of the support he received from Edward D. Libbey. "It is doubtful, however, whether he [Owens] would have accomplished what he actually did if he had worked with someone else. Libbey and Owens made an excellent, harmonious team for what they originally set out to do."[44] Scoville said Owens found it difficult, as the company grew, to associate with the business and finance men who became executives of the company, and Owens did not feel free to be an inventor or entrepreneur anymore.

Between its founding in 1904 and 1920, the net worth of the Owens Bottle Company increased from just over $3 million to over $30 million. But the Owens Bottle Machine was not to be Michael Owens's only revolution in the glass industry. Having conquered the production of bottles, Owens turned his attention to improving the production of flat glass, and with it, started an entirely new revolution and a new company.

In 1983, the American Society of Mechanical Engineers designated the Owens "AR" bottle machine as an international historic engineering landmark. In its proclamation, the organization said "Mike Owens's invention of the automatic bottle-making machine in 1903 was the most significant advance in glass production in over 2000 years."[45]

FOUR

A Second Revolution

> The glass industry of the United States and the world is in another furor similar to that caused by the invention of the Owens machine for making bottles automatically. . . . The machine which has attracted the attention of all the glass manufacturers of the world is designed to manufacture window glass by a continuous sheet drawing process, and it is claimed the invention a success.
>
> —*Toledo Daily Blade*, September 20, 1906, on the continuous sheet glass machine invented by Irving Colburn

Having revolutionized the bottle industry with his automatic bottle machine, Michael Owens sought to do the same for the flat glass industry. He was intrigued by a process still in the experimental stage developed by inventor Irving Colburn. Colburn's process involved drawing a continuous sheet of glass from a melting pot over and through a series of rollers, producing glass of even thickness and constant width with no distortions. Colburn's machine had gained much attention—prematurely, as it turned out—through articles in *Scientific American* and in newspapers around the country in 1906.

At the time Colburn was working on his machine at the beginning of the twentieth century, flat glass was still being made using methods employed in the seventeen century.[1] One method, called "crown glass," involved a skilled glassblower gathering a gob of glass at the end of a blowpipe as he would to produce glass vessels. The blower produced a large spherical bubble, and an iron "punty" rod was attached to the sphere at the opposite side of the blow-

Irving Colburn (in light-colored clothing) with workers at the Libbey-Owens Sheet Glass Company, ca. 1914. (Ward M. Canaday Center for Special Collections.)

pipe. The sphere was broken off the blowpipe, leaving a hole. The glass was reheated, and the punty rapidly rotated to produce a flattened piece of glass through centrifugal force. Crown glass had to used in small windowpanes because of the distortions produced in the flattening process.

Another method involved blowing large cylinders of glass off a blowpipe. The cylinders were then reheated, split open lengthwise, and flattened while still malleable. Again, this process produced distortions because the outside of the cylinder was larger than the inside and stretching the inside to be the same size as the outside produced waves in the glass, even when done by highly skilled blowers. But this process could be used to produce larger panes of glass since the distortion was less obvious than those in crown glass. In 1896, John H. Lubbers invented a machine to produce the large cylinders.

Distortion-free glass could only be made using the plate glass production method. In this process, molten glass was poured onto a flat iron casting table, flowing out into a large rough sheet. After days of cooling in an annealing lehr, it was moved to a large grinder. Fine sand and felt pads lubricated by water ground and polished the glass until it was of uniform width, clear, and undistorted. Glassmakers could produce larger sheets of glass and thicker glass using this method. But plate glass was expensive to produce, and having plate glass windows in your home was a sign of wealth.

Because of the expense of plate glass and the distortions in flat glass, Michael Owens was attracted to the method under development by Irving Colburn. Colburn's method had the possibility of producing large, mostly distortion-free flat glass mechanically and cheaply. But while the method seemed simple, the details were devilish. And as Owens found out, extremely expensive to perfect.

Captain John B. Ford and the Founding of America's Plate Glass Industry

A decade before Owens became interested in flat glass, Edward Ford had built a large factory in Rossford, just south of Toledo, to produce plate glass. Ford was a member of the first family of American plate glass. His father, Captain John B. Ford, founded several early plate glass factories in the United States. Most important among them was Pittsburgh Plate Glass.

John Baptiste Ford, whose "captain" title was acquired during the time he spent building steamboats, was born in Danville, Kentucky, in 1811. He was the third child of Jonathon Ford and his younger wife, Margaret, who married when she was just 15. Jonathon enlisted in the army in 1812, leaving his family to fight against the British at the Battle of New Orleans. He was never heard from again, and Margaret became a widow at the age of 22.

As was customary for children of poor families of the time, John Ford became an apprentice at a young age. But he was unhappy working for a saddle maker in Danville and ran away. He found work with another saddle maker in Greenville, Indiana. He was so successful in this job that he ended up buying the store. His company landed a contract to make saddles for the Mexican War, which made him a wealthy man. He married Mary Bower, who taught him to read and write.

In 1854, he sold the saddle business and opened the Ohio Falls and New Albany Rolling Mills in New Albany, Indiana. From this business he moved into shipbuilding, providing ships to the Union army during the Civil War. He also started a glass company with his son Emory, making bottles and fruit jars.

After a series of successes and failures in numerous enterprises in the rocky economic times following the war, Ford became interested in plate glass manufacturing in 1868. The production of plate glass had been tried and had failed at numerous plants in New England and New York. Despite the industry's track record, Ford built a plate glass factory in New Albany in 1869. "The word fail is not in Capt. Ford's dictionary, and this enterprise cannot fail of success," proclaimed an article about the new factory published in the magazine *Scientific American.*[2] Ford's factory, which opened in 1870, was the first commercially successful polished plate glass factory in the United States.

But competition from the long-established European plate glass industry that could produce plate glass much more cheaply—and a lack of tariff protection—led Ford to close the New Albany plant. He believed a larger factory might be able to produce enough volume of glass to lower production costs to the point where he could compete with European producers. In 1882, he built that factory at Creighton, Pennsylvania. Ford brought his son, Edward, into the business. The company was named Pittsburgh Plate Glass.

Edward Ford was born in 1843 in Greenville, Indiana. When young, he worked as a bookkeeper at his father's steamboat office. In 1864, he married Evelyn Carter Penn. But she died seven years later, leaving Ford a widower with a six-year-old daughter and a four-year-old son. In 1872, he married Caroline Ross from Zanesville, Ohio.

Pittsburgh Plate Glass was managed by the Fords and John Pitcairn, with Edward Ford as president. The younger Ford was constantly trying to improve production techniques. In 1891, he began to use natural gas in glass production. In 1893, he traveled with Pitcairn to Europe to study the more advanced plate glass business there. The two toured the British plate glass factory of the Pilkington brothers, who were pioneers in the field. In an excerpt from a diary Edward kept of the trip, he noted the Pilkingtons' newly renovated facilities and advanced production methods. "The Pilkington Brothers are the most enterprising Plate Glass men, and the most progressive we have met in Europe, and I think they are in the lead," Ford noted.[3]

The Fords and Pitcairn differed in management philosophy. The Fords believed the best way to sell their product was through independent jobbers, while Pitcairn preferred company-owned distributorships. In 1896, Edward Ford resigned as president of Pittsburgh Plate Glass, and he and Captain Ford moved to Wyandott, Michigan, where they opened the Michigan Alkali Company. The company produced soda ash for the glass industry.

But Edward Ford remained interested in plate glass production. In 1898, he sought to purchase a site for a new factory just south of Toledo, where

Edward Ford, founder of the Ford Plate Glass Company in Rossford, ca. 1900. (Ward M. Canaday Center for Special Collections.)

glass manufacturing had been established a decade before by Edward Drummond Libbey.[4] Ford purchased 173 acres along the Maumee River for his factory. The company, named the Edward Ford Plate Glass Company, produced its first successful plate glass in October 1899. Within a year, it was the largest plate glass company in the United States, with the capacity to produce six million square feet of glass. One month after the company opened, Captain Ford visited his son's factory to celebrate the elder's 88th birthday. He would live to be 92 years old.

The factory produced rough plate glass blanks that were transferred from the annealing lehr to an iron grinding table 24 feet in diameter. After the blank was ground on one side, it was turned over and ground on the second side, then polished smooth. In 1907, the company began using electricity for power. In 1911, a second plant opened on the site. The clay pots that held the molten glass doubled in size to hold a ton of glass. The casting tables were enlarged, as were the grinding tables, which measured 36 feet across. Ford hired David H. Goodwillie, a college-educated engineer, who brought scientific production methods into the company, replacing the industry's traditional dependence on the unscientific expertise of experienced glass men.

But Ford did not just build a factory, he also built a town. He named it Rossford, a combination of his wife's maiden name and his own. He sought to provide his workers with a good place to live. In 1899, he donated land for the town's first school. Beginning in 1903, he began building houses and apartments for his workers, which he rented out to them. But there was one business he would not sell land for—a saloon. "Every man who bought property owned by Edward Ford signed a deed that included a clause prohibiting forever the sale of intoxicating liquor on the premises."[5]

The Colburn Process: "Simplicity Itself"

At the same time Edward Ford was expanding and improving on methods for producing plate glass, Irving Colburn was desperately trying to perfect his process of continuous sheet glass production to make commercially viable glass. And his investors were getting nervous.

Irving Colburn was born in Fitchburg, Massachusetts, in 1861. He was the son of an engineer; his father, Henry Colburn, ran several industrial factories in Massachusetts before moving to Toledo in 1890 along with two of his sons to become superintendent of the Herbert Baker Foundry and Machine Works. Irving Colburn stayed in Fitchburg, and experimented with electricity. He installed the first telephone and the first electric lights in Fitchburg, and also manufactured electric motors. Irving's interest in glass production was a result of his brothers' involvement in Toledo with Michael Owens. Both Henry and Leslie Colburn worked on Owens's automatic machine to produce glass tumblers. In 1897, the Toledo Glass Company purchased their patents on this machine.[6]

Irving Colburn thought there must be a way to automate the production of sheet glass. While the machine invented by John Lubbers to produce flat glass from cylinders showed promise in speeding up production, Colburn felt it was a step in the wrong direction because producing flat glass by first blowing a cylinder was inefficient. "I became interested in glass, twelve years ago, visiting factories where window glass was being made by hand, and witnessing the crude methods employed, practically no improvement having been made in generations in the process, and thought it would be a comparatively easy matter to invent a machine that would make transparent sheets of glass," Colburn noted in 1910.[7]

While his idea for drawing a continuous sheet of glass seemed simple, it was frustrating. Colburn was inspired in this machine's design by watching the process used to produce paper, where fibers were pulled over and through rollers into a continuous sheet. Colburn experimented with a glass machine built on the same principle at a factory in Frankford, Pennsylvania, where in 1899 he succeeded in making the first machine of its kind in the world. But the glass produced through the continuous method was rough and not commercially sellable. In 1902, Colburn moved to a larger factory in Franklin, Pennsylvania, and continued his experiments. In 1904, he reported that he had successfully created a machine to draw glass in a continuous width and constant thickness, an improvement over his 1902 model.

While the 1904 experiments were not perfect, two years later he was successful in getting major investors to back his company, thanks largely to the articles about it that appeared in publications nationwide.[8] He named the company the Colburn Machine Glass Company, and it was capitalized at $500,000.

Press accounts of Colburn's invention were glowing. Colburn was granted over 30 patents on the process, and he envisioned plants all over the United States and Europe producing glass using his machine. An article in the *Toledo Daily Blade* in September 1906 reported that the machine had gone through various iterations to work out distortion, brittleness, width, and thickness issues before being perfected. "The final solution, it is said, is simplicity itself."[9] The Colburn machine was compared favorably to the Owens Bottle Machine. "It is said that even the wonderful Owens machine will not do more to revolutionize the business than will this marvel of ingenuity."[10] More investors bought into the hype despite the fact that Colburn's machine had yet to produce any sellable glass.

Between 1905 and 1912, Colburn built and tore down 15 different continuous sheet glass machines. In 1908, a machine was finally installed at the Star Glass Company in Reynoldsville, Pennsylvania, and production began there in 1909. The contract with the company called for Star Glass to operate the machine for 60 days, and if unsuccessful at producing quality glass, the machine would be taken over by the Colburn company at Colburn's expense. While the machine did produce 6,000 boxes of sellable quality glass, it remained unreliable.

In a letter to stockholders of the Colburn Machine Glass Company in August 1910, the directors of the fledgling company indicated the company was in debt $23,000 to repair the Star company machine.[11] The directors added that it would take considerable new investment to perfect the machine beyond what was needed to cover the debt. The letter painted a dire picture. "Unless the present stockholders step in and meet this indebtedness and furnish further capital, in all probability steps will be taken to put the company into the hands of a receiver or in bankruptcy, and the patents and other assets will be sold."[12] The letter made one final plea for stockholders to purchase additional stock.

Owens and a Second Glass Revolution

Michael Owens toured Colburn's factory as Colburn worked to perfect his machine. The two developed a friendship. With Colburn's patents soon to be on the auction block, on May 24, 1911, Owens presented the board of directors of the Toledo Glass Company, the same company that had helped to underwrite Owens's bottle machine, with Colburn's engineering ideas. Owens asked the directors to purchase the Colburn patents at auction. According to an oral history interview recorded years later with John Biggers, Owens threatened to pull out of the Toledo Glass Company and start his own company if the board would not support the purchase of the patents.[13] On February 8, 1912, a representative of the Toledo Glass Company successfully bid $15,000 for Colburn's patents. Most in the glass business thought the patents useless, but a report in the *National Glass Budget* expressed the opinion that the Toledo company would not have taken on such a risk if its managers were not convinced of the ultimate success of the machine.[14]

Toledo Glass gave Owens six and a half acres of land on Castle Boulevard near the Willys-Overland factory to build a plant to perfect the CX—or Colburn Experimental—machine. Over $250,000 was spent on building the plant alone. Owens invited Colburn to help work on the machine, but they were unable to make it a success. In December 1914, Irving Colburn wrote to his brother George that Toledo Glass had already spent $450,000 on his machine, and they estimated it would take another $200,000 before it was completed.[15]

According to Arthur Fowle, later vice president of the Libbey-Owens Sheet Glass Company, few involved thought the machine would ever work. "Practically all others who were privileged to witness the two years of experiments at Toledo were not backward in stating that Mr. Libbey, Mr. Owens, and their associates were foolish in spending such sums of money on the much maligned Colburn machine."[16] Fowle thought it was Libbey who persevered, but it was Owens who was deeply involved in the engineering to work out the technological problems. By May 1915, the machine was finally producing sellable glass. By 1916, the cost to the Toledo Glass Company was well over $1 million.

Libbey and Owens: A New Partnership

After Owens finally achieved success with the Colburn machine, the Toledo Glass Company spun off the operation, as it had done previously with the Owens Bottle Machine Company. The new company, incorporated on May 18, 1916, was called the Libbey-Owens Sheet Glass Company (no doubt an effort to build on the names of the country's two most established men in the glass industry). The new company expanded beyond Toledo, constructing a new $6 million plant in Kanawah City, West Virginia, near Charleston. While Owens's bottle machine company was a large, highly regarded one, the sheet glass company was initially much smaller.

Colburn was granted stock in the company to repay him for his development efforts and to help his family overcome significant debt. Tragically, Colburn would not live to see the full commercial success of his machine. During a trip back home to Fitchburg, Massachusetts, in the spring of 1917, he became ill and was confined to a bed in a hotel for nine weeks before being transported back to Toledo in a special railroad car. He died on September 4. The first unit of the Kanawah plant went into operation the following month.

By 1919, it was clear that the Libbey-Owens Sheet Glass Company had

Edward Drummond Libbey. (Ward M. Canaday Center for Special Collections.)

turned a corner. The annual report to stockholders noted the company had finally begun to realize the potential of the machine that Colburn had touted over a decade before. "The introduction and development of the Colburn patented process of manufacturing sheet glass, in which your company has invested its capital, has now reached the stage where it has proved a commercial success. This could not have been said just two years ago. . . . The almost insurmountable difficulties your management has experienced since the formation of your company in 1916, have to a great extent been overcome and during the greater part of the last fiscal year results have been more encouraging," the report stated.[17] Profits for that year were over $390,000.

While not overwhelming, the profit the company produced was enough for the Libbey-Owens board of directors to expand the Charleston plant. Six machines were added at a cost of $2.5 million. The company also began selling and licensing machines to other companies. In 1920, three machines were sold to the United States Window Glass Company in Shreveport, Louisiana. Three years later, Libbey-Owens took control of that company. Investors in Japan purchased three machines in 1920, and together with Libbey-Owens Sheet Glass created the American-Japan Sheet Glass Company (which later became Nippon Glass). Two machines were installed in Canada, 12 in Belgium, and negotiations were under way with glass manufacturers in Spain, Switzerland, and France. In four years, Libbey-Owens Sheet Glass had

Michael J. Owens. (Ward M. Canaday Center for Special Collections.)

become an international company. Profits for 1920 were reported at $4.2 million. In 1921, the company installed a new machine capable of producing glass up to 84 inches wide.

With success finally at hand for the company bearing their names, Edward Drummond Libbey and Michael Owens began to pull back from daily involvement with both the sheet glass company and the bottle company. James Blair took over both companies as vice president and general manager, with John Biggers, who would play a pivotal role in the sheet glass company's growth for the next 40 years, becoming vice president and assistant general manager. The management between the sheet glass company and the bottle company remained close.

Automobiles and a New Market for Glass

While the Colburn process was fine for windows used in buildings, it still produced too many distortions for moving automobiles, where each distortion was magnified. The only way to produce glass for the booming automobile industry was through the plate glass process of grinding and polishing. Owens pressed Libbey to build a new factory that could be used to produce plate glass for automobiles.

The new factory was to be built on Toledo's east side in 1923 at a cost of $2.5 million. Owens also wanted to experiment with laminated glass, which was safer for autos because it would not shatter into dangerous shards in an accident. He outlined the value of safety glass in an interview that appeared in the popular *American Magazine* in July 1922.[18] "Suppose you were in a closed automobile during an accident. Or suppose you were sitting in the front seat of any motor car with the glass windshield in front of you. Scores of persons have been terribly cut by broken glass under such circumstances. It is not an uncommon thing for a person to be thrown against one of the windows with force enough to break it, and to receive dangerous cuts around the face and head," Owens said. "Sometimes a person's hand and arm go through a window and are badly cut," he added. Owens then demonstrated for the interviewer what happened when the safety glass was hit. "But this [safety] glass does not shatter in pieces. If you strike it hard enough, it shows these minute cracks," Owens said.

As with his bottle machine company, Owens was a demanding manager of the sheet glass company engineers. Hints to his demeanor are revealed in a notebook that was kept by Joseph Crowley and E. G. Peters recording the experiments carried out by the engineering department of the Libbey-Owens Sheet Glass Company. The book noted many visits Owens made to the department while the engineers struggled to perfect the grinding and polishing process that would be used in the new East Toledo plant, and conveyed a sense of the demands Owens made of them.[19] On July 8, 1921, Peters wrote, "Saw Mr. Owens and Mr. Blair in Mr. Owens office and they are anxious to get LO 1 [Libbey-Owens 1] machine running." Further entries included: "Joe calls M. J. Owens over the phone and was told to keep the machine going just as it is for Mr. Libbey's inspection" (June 6, 1922); "Owens told Crowley to design a new grinder and new polisher to take care of glass large enough to be cut in 3 windshields" (June 22, 1922). Some engineers were abruptly fired by Owens. "Joe [Crowley] announces that F. Stroder has been discharged immediately by M. Owens. He left at 12:00 noon," the volume noted on July 3, 1922.

Despite the expense of developing a new grinding and polishing process and building a new plant, profits for 1922 were $3.5 million. That year, Libbey-Owens Sheet Glass also built 49 houses and 42 apartments in Charleston for the workers at the plant there, which the company hoped would improve not only their standard of living, but also company loyalty.[20]

But all was not rosy. That same year the company had to close a factory in Hamilton, Ontario, due to high production costs, low demand, and high tariffs, which it never reopened. The company was pouring millions into research and development, especially in the production of laminated safety glass for the auto industry. Glass was in such a demand by the big auto producers that in 1920, Henry Ford tried to buy the Edward Ford Plate Glass Company in order to procure a constant supply of windshields. Libbey and Owens had to move quickly if they were to meet the demand for auto glass.

Death "in the Harness"

In addition to his work for the Libbey-Owens Sheet Glass Company, Michael Owens continued to be involved with the bottle company that bore his name, but less with day-to-day matters. In the 1922 *American Magazine* interview, he stated he was still working on improvements to his bottle machine.[21] The article's author noted that Owens had a blackboard in his office covered with figures and the phrase "Do not erase" written at the bottom. Then 62 years old, Owens expressed his belief that hard work was key to success. "Young or old, work doesn't hurt anybody," Owens said.[22] When asked if he intended to retire, Owens said he wanted to continue to improve the company and keep it competitive. "The real reason I keep on is because I like to, I want to work. It is the most interesting thing in the world and it is the most constructive thing. I've enjoyed 52 years of it, and I hope to enjoy a good many more," Owens added.[23] That summer, Owens became ill while visiting the East Coast, but he refused to give up his responsibilities with either company.

Owens would not live to see his 65th birthday. On December 27, 1923, while attending a contentious meeting of the directors of the Owens Bottle Company in the company's headquarters in the Nicholas Building in downtown Toledo, Owens suddenly left. Those in attendance did not notice anything peculiar, and gave little thought to his departure. The stories of exactly what happened next as reported in the city's newspapers differed somewhat. One report said Owens went into one of the company offices and announced that he was feeling ill and asked that a physician be called. Another said he was walking down the corridor when he grabbed his chest, and he was helped into a chair. Two of his doctors were summoned, as was his priest, Father A. J. Dean. He was pronounced dead 25 minutes after leaving the meeting. He left behind a wife and a grown son and daughter. The meeting of the board of directors quickly moved to the business of planning a memorial service.

At his other company, the news was equally devastating. The board of directors of the Libbey-Owens Sheet Glass Company called a special meeting two days later, and a resolution was entered into the minutes of the company in honor of Michael Owens. It read, in part: "Those among us who have had the good fortune to be associated with him during the stress and strain of the day's work know that the courage, vision and perseverance of the Genius were no greater than the sterling qualities of the Man. Truly those attributes which enabled him to revolutionize a vast industry were no greater than the magnetism and charm which endeared him to us. The timepiece of Eternity struck for him and called him as he had wished—engaged in the full exercise of his extraordinary talents in his lifelong work, and, as his memory melts into the shadows of the past, he will always have an exalted place in the hearts of his many friends who are carrying on the work which he so broadly conceived."[24]

While the true nature of the relationship between Owens and Edward Drummond Libbey will never be known, on Owens's death Libbey issued a heartfelt tribute that was published on the front page of the *Toledo Times*. "Self educated as he was, a student in the process of inventions with an unusual logical ability, endowed with a keen sense of farsightedness and vision, Mr. Owens is to be classed as one of the greatest inventors this country has ever known. He had done more to advance the art of glass manufacturing than any other person during the last 50 years. . . . I believe that the name of Michael J. Owens will stand out as a pronounced example of what can be accomplished by vision, faith, persistence and confidence in one's creative efforts," Libbey wrote.[25] He added, "He died as he wished, in the harness, without relinquishing his duties and among his business associates." Other tributes followed. The *Toledo Blade* noted that "Toledo is a better and busier and a richer town because he lived here and posterity will owe him much for revolutionizing a great industry."[26]

While Owens invented the machine that forced many skilled glassblowers out of their jobs, he was still revered by the glassworker unions, who saw him as one of their own. The AFGWU paid homage to Owens at the time of his death. "Mr. Owens had accumulated millions, but his attitude toward his fellow-men did not change—he was still the same 'Mike' Owens, and the humblest person could feel at home in his presence. He was a philanthropist, and many of those less fortunate have a just cause to mourn because he is no more. A noble soul departed when the genius of the glass industry—Michael J. Owens—died," an article in the union's magazine, *American Flint*, stated.[27] Five years after his death, when the union held its convention in Toledo, he was still on the minds of members. "Michael J. Owens was one of the most earnest, aggressive, and fearless men holding membership in our union. Many of the accomplishments of the union could be credited to him. We can turn the pages of our history and find that he left his impress upon them, and we can testify that while he was active in the affairs of the union he brought to his task a clear mind and a determination worthy of emulation," the union's magazine noted.[28]

The year of Owens's death, the three companies tied to him and Libbey—Libbey Glass, Owens Bottle, and Libbey-Owens Sheet Glass—were separated in their management. This helped to clarify responsibilities. Work continued on the East Toledo plate glass plant, which was still in its experimental stage. It finally opened on July 1, 1925.

The changes that had occurred in the glass industry since 1899, when Edward Ford opened his plate glass factory in Rossford, had been profound. In 1899, America's glass industry produced $17 million worth of goods. By 1925, that had risen to $72 million. This dramatic increase had occurred by adding only 17,000 workers nationwide—from 52,000 employees in the industry in 1899 to 69,000 in 1925.[29] Clearly, the automation introduced by Michael Owens had impacted glass manufacturing.

Death Claims the King of Glass

Following the death of Michael Owens, Edward Drummond Libbey spent more time away from all of his companies, and more time traveling extensively to collect art or relaxing at his estate in California. In May 1925, he resigned from the board of directors of the Owens Bottle Company. He was still somewhat involved in the Libbey-Owens Sheet Glass Company, taking a contingent of European glass officials to the Charleston, West Virginia, plant in September 1925. But it was his civic and philanthropic interests that occupied most of his time.

Libbey had served on the Toledo Board of Education from 1911 to 1914, and had established several scholarships for Toledo children. In 1922, the Toledo Board of Education decided to name a new high school planned for the city's south side in Libbey's honor. Libbey was flattered, and donated $35,000 for an athletic field for the school. The school was officially dedicated as the Edward Drummond Libbey High School on December 6, 1923.

Libbey spent much of 1924 in Europe and the summer of 1925 in Egypt, and was planning to visit his estate in Ojai, California, for several weeks in the fall. As in Toledo, Libbey was involved in civic affairs in Ojai, including helping to design the town with Spanish mission-inspired architecture. In Toledo, a new expansion of the Toledo Museum of Art that Libbey donated $850,000 to construct had begun in 1924, and was ready to be occupied in the fall of 1925.

But in the first weeks of November 1925, he became ill at his residence in the Secor Hotel in downtown Toledo. He developed pneumonia, which was complicated by chronic heart disease. Despite care from the best doctors in the city, he died at the hotel on November 13, just two years after the death of Michael Owens. In its headlines announcing his demise, the *Toledo Blade* proclaimed, "Death Claims E. D. Libbey, Glass King."[30]

The outpouring of grief over Libbey's death overwhelmed the city. Because of the reach of his activities and his philanthropy, he had touched many beyond his wealthy business partners and aristocrats. All of the public and parochial schools in the city held a memorial service, and a special memorial service was held at the high school that bore his name. The football game between Libbey and Woodward high schools was postponed. Many organizations passed formal resolutions expressing their condolences. One particularly moving tribute was by employees of Toledo Hospital who, upon hearing of the death of Libbey, pushed the fund-raising campaign for a new hospital over the $500,000 mark on the inspiration that Libbey had provided.[31]

The Reverend Allen Stockdale of First Congregational Church, who conducted Libbey's memorial service, said, "We are all quieted by the loss of a great man. We needed him; we wanted him; we loved him. His vision must be picked up by some following mind. His heart impulses must be accepted by those who love him. We are humble; we are reverent; we are sad. Help us be true to the vision made clear by one who has gone. Help us to love our city by every good work and deed."[32]

The expressions of sadness were most noted by the companies he helped to create. The directors of the Owens Bottle Company stated, "We have not lost a friend, we have but discontinued for the present his association, and his inspiration and counsel of former years will remain our guide, our continued effort to make our company one of the many, many monuments to his life."[33] At a special meeting of the board of directors of the Libbey-Owens Sheet Glass Company held five days after Libbey's death, the board unanimously passed a resolution of tribute. "[Libbey] was endowed with great vision and enobled by high ideals and inspiring character. To know him was to love him. His energy was boundless, his enthusiasm that of perpetual youth. . . . His good works are many, and eloquently speak in enduring tribute. However, his greatest deeds were wrought in the lives and the hearts of untold numbers who were thrilled by his noble example and who were quickened to greater efforts and to greater achievements through his inspiring leadership and unceasing help."[34] In his comments at the annual meeting of Libbey-Owens Sheet Glass stockholders, James C. Blair, vice president of the company, spoke of the responsibility owed by the entire company to continue Libbey's legacy. "He has left the management and employees of this company an important responsibility—a responsibility not only to the owners of the company, but to every one in this community, on account of the large public bequest he made of his material possessions, a substantial part of the value of which is dependent upon the prosperity of this company."[35] The company's advertising in 1927 drew connections between the company founders and the great inventors throughout history. "Edison in Electricity. Bessemer in Steel. Marconi in Wireless. Libbey-Owens in Glass."[36]

After his coffin lay in state in the newly completed wing of the Toledo Museum of Art, Libbey was buried on November 16, 1925, at Woodlawn Cemetery in west Toledo. His estate was valued at $21 million.[37] It was the largest estate ever recorded in Lucas County. It included nearly $3 million in stock in the Owens Bottle Company and $2 million in the Libbey-Owens Sheet Glass Company.

But in many ways his contributions to the city only began with his death, as nearly half of his estate ($10 million) was designated for charity.[38] The estate's largest donation was to the museum that he and his wife founded, which received $4.58 million. Included in this money was funding for building a new venue for musical entertainment and public lectures. In 1931, at the depths of the Depression, that money was used to construct the Peristyle and the School of Design wing of the Toledo Museum of Art, a project that helped to employ some 3,000 men for two years. Other funds went to the Toledo Public Schools, to the public library, and to a trust established to carry on Libbey's philanthropy. Many local charities benefited from the trust for years after Libbey's death.

Two Companies Minus Their Founders

After his death, the Libbey estate shares in Libbey-Owens Sheet Glass were offered for sale. Raymond, Joseph, and Robert Graham, three brothers who had long been involved in the glass and automotive industries, successfully bid on the shares along with an investor group that included retailing magnate Marshall Field, the investment bank Lehman Brothers, and a Belgium glass interest. By 1928, the Grahams owned 3,200 shares, and all three brothers joined the Libbey-Owens board of directors.

The Grahams got their start in the glass industry in 1908 when their father founded a small bottle factory in Indiana. It evolved into the Graham Glass Company and expanded to three factories in Indiana and Oklahoma. Ray Graham was secretary and treasurer of that company. In 1916, the Graham Glass Company was sold to the Owens Bottle Company, the first liaison of the Grahams with the Toledo glass industry. The Grahams also were interested in truck production, and in 1917 created the Graham Truck Company. Their truck company was eventually sold to Chrysler Motor Company, and it was profits from this sale that allowed the brother to invest in Libbey-Owens Sheet Glass stock. The brothers continued an interest in automobiles, buying the Paige-Detroit Motor Car Company in 1927.

Because the Grahams were also involved in automobile manufacturing, they were particularly interested in expanding production of laminated safety glass. Many automobile manufacturers like Ford were adopting safety glass for their cars. Laminated glass was put into experimental production at Libbey-Owens's East Toledo plant in an effort to develop a product that could meet the demand. A report prepared for the board of directors in 1929 showed the dramatic growth in demand for plate glass by the automobile industry in the five previous years.[39] In 1924, the industry used 41.2 million square feet of plate glass. Just five years later, the demand had more than doubled to 99.9 million square feet. The percentage of plate glass used in

The grinding and polishing lines at the Ford Plate Glass Company. (Ward M. Canaday Center for Special Collections.)

automobiles had risen from 39 percent of all plate glass produced in the United States to 63 percent. The market for plate glass, and particularly laminated safety glass, seemed nearly unlimited, and the future of the product was clearly tied to the automobile.

The Libbey-Owens Sheet Glass Company struggled without its founders, and so too did its chief competitor, the Edward Ford Plate Glass Company. Edward Ford died on June 14, 1920, and was succeeded by his son, George Ross Ford. But George had little interest in the glass industry. The Ford company was also attempting to expand and improve on production methods, and in 1928 paid $25,000 for rights to use the Bicheroux process. The Bicheroux machine used water-cooled iron rollers to control the thickness of glass as it was cast, thus reducing the grinding and polishing required. The new process sped up production to the point that a continuous grinding and polishing line was required, which was installed in 1930 by technicians from Belgium. The new line meant the end of the 36-inch grinders that had been in operation for 15 years. All of the improvements in engineering had drastically increased the capacity of the Rossford plant since its founding. In 1900, the year after the factory opened, it produced 1.1 million square feet of plate glass. By 1929, it was producing 14.2 million square feet.

So as the decade of the 1920s ended, northwest Ohio had two companies producing plate glass, both of which had lost their founders within five years of each other. The Edward Ford Plate Glass Company had faster production capacity for large plate glass through its Bicheroux line. The Libbey-Owens Sheet Glass Company was fairly new to plate glass, but it was ahead of the Ford company in its experimentation with laminated glass. While plate glass would always be necessary for the construction industry for architectural windows, laminated safety glass for the automobile industry was where future growth was aimed. A merger would bring together complementary technology and strengthen both companies. And a merger would also bring together the three great names of Toledo glass.

The Merger

In 1929, Ray Graham became chairman of the board of the Libbey-Owens Glass Company. He replaced James C. Blair, who had been Edward Drummond Libbey's right-hand man in the management of both the bottle company and the sheet glass company. Blair had overseen the rapid expansion of Libbey-Owens from 1916 to 1930, and the company dropped the word "sheet" from its name to reflect its expanded product offerings. On February 10, 1930, Graham and others from Libbey-Owens met with George MacNichol Jr., W. W. Knight, and John Ford (George Ford's brother) from the Edward Ford Plate Glass Company, in Miami, Florida, to hammer out details of a merger. In a letter to shareholders of Libbey-Owens dated April 30, 1930, the terms of the merger were explained—as well as the reasons for it. "Its [Edward Ford Plate Glass] products and methods have earned for it, under the capable management of three generations of the Ford family, a good-will and esteem of inestimable value. Its physical properties will supplement ours without wasteful duplication."[40] The letter also noted that the Ford company had adopted the more efficient Bicheroux process the year before, while rights to that process were owned by Libbey-Owens in a collaboration with other flat glass producers worldwide, another reason supporting the merger. The letter called for a special shareholders meeting of Libbey-Owens to be held on May 20.

A similar letter went out to the shareholders of the Edward Ford Plate Glass Company on May 12, 1930.[41] It was explained that for each share of the Ford company owned by the stockholders, they would receive 4.75 shares of the new company. The directors expressed their belief that the merger was advantageous to shareholders, and unanimously recommended its approval. A meeting of the shareholders was called. The board of directors of the Ford company demanded that the name of the new company reflect the name of its company as well, and "the Libbey-Owens-Ford Company" was agreed upon. The stockholders of both companies approved the merger, with the final vote taken by those of Libbey-Owens on May 26, 1930. With the merger, the capital value of the company increased from $11.5 million to $14.5 million, and the total assets increased from $30.8 million to $44.5 million.

But the merger came at a difficult time for the glass-manufacturing industry, as the devastating impact of the Depression began to be felt. In just two years, from 1928 to 1930, residential construction had declined 57 percent.[42] A decline in housing construction meant a similar decline in window sales. In addition, automobile production, another industry that glass manufacturers depended on, declined 30 percent over the same time period. While the newly formed company had much to celebrate, it also had much to contemplate as it struggled against the country's economic headwind.

The Depression and the New Company

The new Libbey-Owens-Ford Company (L-O-F) appointed a new president in 1930. With the exception of four years that he worked in the automobile industry, John D. Biggers had worked nearly his entire life in Toledo's glass industry. He came to Toledo in 1911 at the age of 23 as secretary to the Chamber of Commerce. But he was soon recruited away from that position by William Walbridge and became assistant treasurer for the Owens Bottle Machine Company, working for Michael Owens. He later became assistant general manager of the company. In 1926, he was hired by the Graham brothers to increase auto and truck sales at the Dodge Brothers British Company, and Biggers moved to England. With the creation of the new Libbey-Owens-Ford Company, he returned to Toledo and became president, a position he would hold for the next 30 years.

As president, Biggers was faced with the continued decline in sales of L-O-F products during the Depression. He felt the automobile industry

John Biggers, who became president of Libbey-Owens-Ford in 1930. (Ward M. Canaday Center for Special Collections.)

presented the best opportunity for improving the company's bottom line. In 1930, he approached Fred and Charles Fisher, owners of the Fisher Body Company, to inquire if L-O-F might contract with the company to provide windshields for General Motors automobiles.[43] To Biggers's surprise, the Fisher brothers offered instead to sell L-O-F their National Plate Glass Company in Ottawa, Illinois, which would make L-O-F the sole producer of windshields for GM, give L-O-F 45 percent of the country's market for plate glass, and allow the company to become a major competitor to Pittsburgh Plate Glass (PPG).

But there was a catch. Fisher had also made the same offer to PPG—$7.5 million for the plant and the GM contract. Biggers realized that this purchase was a huge risk to the company in the middle of the worst economic depression of the twentieth century. But he also realized this was the opportunity of a lifetime for the new company. He reluctantly agreed to the price. He returned to Toledo and convinced the L-O-F board of directors to agree to the purchase based on the argument that if PPG bought the plant, L-O-F would control just 20 percent of the plate glass market.

In late 1930 and early 1931, Biggers traveled frequently between Toledo and Detroit to work out the deal. But suddenly, Charles Fisher told Biggers that the offer had changed. PPG had upped its offer to $9.5 million, but Fisher said he would allow L-O-F to meet the new offer. In an oral history interview Biggers recorded in 1967 and 1968, he recalled his conversation with the Fishers and how difficult the decision had been for him to make. "Gentlemen, because under these general business and financial conditions, I'm afraid that if I go back to Toledo and this was hashed out by everyone concerned, we might not do it. Yet I am sure, as sure as I am of anything in the world, that this is *the* great opportunity for Libbey-Owens-Ford to rise in importance and stature in the glass industry and we may never have another comparable opportunity, so I will say we'll meet their offer of $9.5 million," Biggers recalled.[44] He believed the reasons the Fishers were willing to sell to L-O-F was both personal and financial. "If they chose to deal with this small, responsible, and honorable company, fully capable of making glass to their requirements, they would create a second strong source in the industry," Biggers said.[45] Biggers was going out on a limb, taking a huge risk in agreeing to the contract. "I tell you quite honestly I don't know where our company will get the additional $2 million, but we will commit ourselves to do so," Biggers said in closing the deal.[46] The L-O-F board of directors approved the purchase at a special meeting on June 15, 1931.

On July 3, 1931, L-O-F sent out a notice to stockholders of the deal. The company issued $9 million in 5 percent gold notes that could be converted to L-O-F stock to finance the purchase. The sale of National Plate Glass to L-O-F left just two companies in control of the entire plate glass market in the United States—L-O-F and PPG.

While the merger of Libbey-Owens Sheet Glass and Edward Ford Plate Glass and the acquisition of the General Motors contract may have saved the company during the Depression, it did not bring immediate prosperity. Wages and salaries of all workers were cut twice by 10 percent, once in 1931 and again in 1932. The company struggled without access to some of its cash that was held in closed Toledo banks. The Ottawa, Illinois, plant that it had purchased just the year before closed in 1932, but the directors promised to reopen it as economic conditions improved. In 1933, the board of directors approved a $75,000 loan to Graham-Page Motors, owned by the Graham brothers, because of the closure of banks in Detroit where the company had its money on deposit.

In the middle of these business difficulties, in 1932 Ray Graham suffered what was reported as a "nervous breakdown."[47] As part of his recovery, he and his brother Robert and a priest who was a family friend were visiting

Chatham, Ontario. After attending mass, Ray leaped from a car driven by his brother into a tributary of the Thames River, and drowned. A resolution passed by the directors of L-O-F detailed the important role Ray Graham had played in the company. "His contributions to the achievement of this company were of great and lasting value. Across the land—other industries, happy workmen are a living tribute to his vision and a monument to his memory."[48] With Ray's death, the management contract that the three brothers had signed with Libbey-Owens Sheet Glass was renegotiated, with Joseph Graham succeeding his brother on the board of directors of L-O-F.

Happy Days Are Here Again

The year 1932 was a bleak one for L-O-F. In addition to Ray Graham's death, the company reported a yearly loss of $295,000. But the company made a dramatic turnaround in 1933, reporting annual profits of $4.2 million.[49] Safety glass production was the reason for the reversal in fortunes, with sales up 43 percent in 1933 over the year before. The product went from 20 percent of the company's total sales to 40 percent in that one year. Window glass also improved. Under provisions of the National Recovery Act, two pay cuts to workers implemented in the previous years were restored and a new minimum wage was enacted. Some 1,200 new employees were added by reducing the workday from eight hours to six hours, which the company did while maintaining wages at the equivalent of an eight-hour day. The Ottawa, Illinois, plant reopened after two years.

Employment at L-O-F had fluctuated dramatically since 1929. That year, the company employed 2,255 workers.[50] But that number fell to less than 2,000 by 1932. With the company's recovery, employment rose to 3,500 in 1933, and by 1935 was up to 6,400 workers.

The company continued to push for more use of safety glass in automobiles. The public also began to demand safety glass, and by 1934 nine states had passed laws requiring safety glass in all windows in the cars sold in those states. While all automobile manufacturers had adopted safety glass for the front windshields by this time, some did not use it in side or back windows. In 1934, General Motors agreed to add safety glass as standard equipment in all of GM cars. To help sell safety glass and increase sales, L-O-F joined with automobile manufactures to promote the product in advertising. The new sales campaign included slogans such as "A car is no safer than the glass in its windows," and "Protection for those Who Ride in the Back."[51] The advertisements featured photographs of cherubic children and the phrase "Counting on You to give them the best available protection against ugly cuts and tragic scars."[52] The ads noted that having safety glass installed in all windows added just $1.50 to the cost of the typical monthly car payment.

Under the leadership of John Biggers, the company sought not only to expand its auto glass business, but its entire product line. Biggers was convinced of increased opportunities for home sales under the new Federal Housing Act, which sought to expand home ownership through attractive financing plans. In a speech before the National Glass Distributors' Association in 1934, Biggers urged the salesmen to push for greater use of windows in the estimated 330,000 new homes that would be built in 1935. "The modern use of windows fuses the outdoors and the indoors in pleasing harmony. Corner windows banish those triangles of shadows that once occurred in every corner of a room. Window walls provide a clear, sunny workspace in the kitchen and other rooms," Biggers told the salesmen.[53] He also promoted the sale of storm windows to save energy costs. He urged the salesmen to work with architects to incorporate more glass into home and building designs, and even urged them to work with decorators to include more mirrors in interior design.

Biggers also took advantage of another federal New Deal program that provided loans to commercial businesses to modernize their buildings. The program, called "Modernize Main Street," provided low-interest loans of up to $50,000 through the Federal Housing Authority for modernizing renovations. In response to the program, L-O-F sponsored a contest with *Architectural Record Magazine* where architects submitted drawings for modernizing food stores, drugstores, automotive sales and service centers, and apparel shops. The jury, which included noted architect Albert Kahn and was chaired by J. Andre Fouilhoux, looked at the way each entry addressed the psychology of selling through the exterior and interiors of the designs—and, of course, how each used glass in its designs. The winners in each category received $1,000, and the winning designs and those receiving honorable mentions were published by L-O-F in a large-format, full-color book entitled *52 Designs to Modernize Main Street with Glass.*

The designs extensively utilized a new L-O-F product line called Vitrolite. Libbey-Owens-Ford purchased the Vitrolite Company in 1935, the same year L-O-F launched its "Modernize Main Street" design contest. Vitrolite was an opaque, colored structural glass product used both on the exterior and in the interior of many buildings built in the Art Deco style of the 1920s and 1930s. It was especially popular on movie theaters, diners, and storefronts. It was also featured prominently in the new Toledo Public Library, opened in 1940, in the children's room and in an extensive mural depicting Toledo's industrial development in the building's main lobby.

The company also purchased the patents to Thermopane in 1934 from the inventor Charles Havens of Milwaukee for $25,000. Thermopane was a product that bonded together two sheets of glass using a metal seal around the outside that reduced the problem of heat loss through windows. But Thermopane took years and millions of dollars to perfect because the seals would break down over time and allow moisture between the double panes. Once perfected, however, Thermopane glass became a hugely successful product for the company, and opened up new architectural design possibilities for office buildings. It was especially important for buildings that were cooled by air conditioning. By 1946, Thermopane was such a large product line that a new plant was built to produce it across from the original plant in Rossford.

Many of the new products were developed in L-O-F's Technical Building located next to the East Toledo plant. Others were developed by other glass companies and purchased by L-O-F. The company's 1937 annual report to stockholders listed 10 major product lines: polished plate glass, window glass, safety glass, Vitrolite structural glass, Tuf-Flex tempered plate glass, Vitrolux colored tempered plate glass, Aklo plate glass (which was chemically coated to absorb infra-red light), bent glass, Extrudalite (glass with metal reinforcement for commercial buildings), and Blue Ridge Figured and Wire Glass (translucent and fire resistant glass).[54]

Labor Unrest

The right of workers to unionize and collectively bargain was supported by the Roosevelt administration through the passage of the Wagner Act in 1933. Many laborers who were barely surviving on the depressed wages created by the economic conditions of the country saw an opportunity to improve working conditions. Toledo quickly developed a reputation as a strong labor union town. This was due in part to the violent Electric Auto-Lite strike of 1934, when 6,000 strikers clashed with Ohio National Guardsmen in a bloody confrontation that lasted five days and resulted in the deaths of two union members and injuries to nearly 200 others. The strike was one of several large strikes that year in the automobile industry that would lead to the founding of the United Automobile Workers.

In 1935, Libbey-Owens-Ford workers joined the newly formed Federation of Flat Glass Workers, which held its first national meeting in Columbus, Ohio, that year. Several glass plants were the subject of strikes in 1935 and 1936. The largest started in October 1936 when 6,000 members of the Federation of Flat Glass Workers struck the Pittsburgh Plate Glass Company, closing all but one of its factories. In December, 1,300 L-O-F workers at the Ottawa, Illinois, plant joined the strike in solidarity with the PPG workers when L-O-F accepted a $4 million contract with Chrysler Motor Company.[55] The L-O-F workers saw the contract as an attempt to break the PPG strike.

Up until the strikes, the plate glass industry had reported record demand and profits. Production was up 20 percent over 1935 levels. But the strike dramatically reduced output. All L-O-F operations were shut down for six weeks and idled 5,850 workers, most of them in Toledo. Between the strike at L-O-F and PPG, 90 percent of the flat glass industry was shuttered. The industry attempted to meet demand by increasing imports from Belgium. Both companies were able to resume domestic production in January 1937 when a new contract granting an eight-cent-an-hour increase in wages was approved.[56]

The Depression Ends and War Begins

While the end of the Flat Glass Workers strike in 1937 was welcomed, it did not end all of L-O-F's difficulties for that year. In June, the Federal Trade Commission filed a complaint of unfair trade practices against the companies that made up the Window Glass Manufacturers Association.[57] The association included the two largest producers of flat glass—Pittsburgh Plate Glass and Libbey-Owens-Ford. The complaint alleged that PPG and L-O-F had embarked on a monopoly beginning in 1928 over the sale and distribution of window glass. The companies admitted most of the allegations, and ended the practices cited in the FTC complaint.

Another of the company's leaders, George Ross Ford, died in 1938. As the son of founder Edward Ford and the grandson of the founder of the plate glass industry in the United States, John B. Ford, George had deep roots in the company. He had become president of the Edward Ford Plate Glass Company in 1920 when his father died, but had retired from the glass industry to become a vice president for several Toledo area banks, including the Rossford Savings Bank. Ford's death marked yet another separation from the company's founders, which the board of directors of L-O-F noted in a resolution passed on George's death: "It is our conviction that this company has, in large measure, acquired its standing in the business world from the fact that George Ross Ford exerted his influence along the lines of his family's tradition—namely, stability, integrity, and fairness in business transactions."[58]

Libbey-Owens-Ford continued its pattern of growth up until the end of the decade. In June 1940, it acquired a controlling interest in Plaskon, a company created from research jointly conducted by Toledo Scale and the Mellon Institute of Pittsburgh that produced urea formaldehyde resins. With this purchase, L-O-F moved outside of glass production into plastics. The company's annual report to stockholders that year highlighted its six factories: Charleston, West Virginia (described as the world's largest window glass plant under one roof); Shreveport, Louisiana; Rossford; East Toledo; Parkersburg, West Virginia (where Vitrolite was produced); and Ottawa, Illinois (which exclusively produced safety glass for automobiles). Having survived the Depression, the company was ready for its next challenge—war.

FIVE

Bottles Everywhere, but Nothing to Drink

> No social function, whether it be a quiet afternoon's bridge or a typical evening party, is complete without being enlivened by serving a good ginger ale or some other carbonated beverage. Their sparkling bubbles lift cares and responsibility and their appealing tang adds zest to food and joy to living. The remarkable growth in the consumption of carbonated beverages is a tribute to the efforts which producers have put forth to improve quality and flavor.
>
> —Owens-Illinois glass container catalog promoting nonalcoholic beverages, 1930

What does a company do when one of the largest markets for its products becomes illegal?

That is exactly what happened to the Owens Bottle Company in 1919 when Congress and the states passed the Volstead Act, which became the Eighteenth Amendment to the U.S. Constitution. The law forbade the manufacture, sale, transport, import, or export of alcohol—alcohol that was sold in bottles made by Owens directly, or by companies using licensed Owens machines.

Prohibition ushered in two difficult decades for the Owens Bottle Company that saw it not only lose a major market, but also its founder. Profits for the company declined from $4.2 million in 1920 to $1.3 million in 1921. The company's annual report that year bluntly spelled out the difficulties. "The year 1921 was a trying one for business generally. So severe and wide-spread was the decrease in buying that the great majority of industrial concerns found it impossible to operate at normal capacity at any time during the year. . . . The bottle industry, still suffering from the adverse effects of prohibition, was particularly affected."[1] Putting the best spin on the company's decline in profits as it could, the report added, "It is an evidence of the fundamental soundness and strength of The Owens Bottle Company that it was able to come through this trying period in excellent financial condition and with substantial profits." The report noted that soft drink sales that the company had been banking on declined and, "coupled with the entire loss of the beer bottle business, reduced operations very materially."[2] The company was operating at a third of its capacity. The only bright spot that year was a large crop of tomatoes that increased demand for catsup bottles.

It was not only the company leaders who were concerned about the impact of Prohibition. So too were the workers who made bottles and other glassware related to the liquor industry. T. R. Rowe, president of the American Flint Glass Workers Union, noted the impact of Prohibition on just one Ohio plant—the bottle factory in Newark. "Previous to the enactment of this farcical prohibition law we employed approximately 1000 men steadily throughout the year. Since 1920 the number on our payroll has dropped to as low as 250 employees and this number was not steadily employed."[3]

In 1923, the death of Michael Owens was yet another blow to the Owens Bottle Company. In 1924, the company's annual report again noted the impact of overcapacity in the industry as a result of Prohibition. Because most of the 120 glass container companies in the United States used Owens machines, in effect the company was competing against itself. Some of the licenses that Owens had granted to other companies had been negotiated to include exclusivity clauses; this meant that Owens was shut out of certain categories of bottle production and could not produce a full line of products even though it had built the production machinery. Royalties paid by the outside companies remained steady, but few could afford to purchase new bottle machines. The company attempted to reassure investors. "It is still impossible to operate your companies' subsidiaries to their full capacity as the loss in business resulting from prohibition has not been made up by the normal growth of the country of any corresponding increase in other lines of ware. Nevertheless, your Company's position in the glass container industry, its well located and modern plants, and its standing with the trade have enabled it to come through the year of only fair business, with increased profits, and point to its greater prosperity when those conditions, which are temporarily adverse, shall have been improved."[4] While profits increased from $3.7 million in 1924 to $5 million in 1925 and $6.9 in 1926, by 1927 they were back to $4.6 million.[5] Plants in Clarksburg, West Virginia, remained idle, and the plants in Toledo and Cincinnati were dismantled.

The Boys from Illinois

In 1928, the Owens Bottle Company was looking for a merger that might serve as a lifeline. Illinois Glass of Alton, Illinois, dated back to 1873, when Edward Levis invested in a glass company founded by William Elliot Smith. Smith's company had already failed twice, and Smith needed an influx of capital to keep it going. Levis agreed to put up some money for the company if Smith would guarantee a job for each of Levis's six sons. Levis invested about $5,000 that he either borrowed from his furniture business or from his retirement savings, while Smith invested $10,000. Smith became president of Illinois Glass, while Edward Levis was vice president and superintendent of operations. The company operated out of a couple of small buildings, and had a five-pot glass furnace. It produced bottles and other glass containers, and with Levis overseeing production and Smith operating the business office, the company became successful.

After the deaths of Edward Levis and his son Edward Levis Jr., son Charles directed the factory's operations. The company continued to expand, making its own molds and shipping boxes and expanding its warehouses. It became one of the largest handblown glass bottle companies in the country, employing hundreds of glassblowers. In 1910, the company risked its reputation by purchasing its first Owens bottle machine, which was only six years into production. But the risk paid off, and by 1915 all of the glassblowers had been replaced by the machines. Illinois Glass became the largest maker of bottles in the world, with a second factory in Gas City, Indiana, offices in St. Louis and Chicago, and part ownership in a West Coast firm located in San Francisco. A sales convention program from the 1920s included the words to a song that summed up the early years of the company: "The Levis boys they built a mill, they built it on the side of a hill, they worked all night and they worked all day, to make the goddamned glasshouse pay."[6]

The company's logo was the letter "I" within a diamond shape. The slogan remained from its days of handblown bottles: "Bottles of Every Description."

Like the Owens Bottle Company, Illinois Glass was adversely impacted by Prohibition. Although making whiskey bottles was still legal, demand was down. Most were sold to pharmaceutical companies (for "medicinal whiskey") or to Canadian companies. To compensate for low demand for liquor bottles, the company turned to the soft drink business. An article in the company publication *Bottles* in 1919 titled "Soft Drink Prosperity" noted that some alcoholic beverage producers had successfully switched to soft drink production.[7] But a year later, the company admitted that soft drinks still had a way to go and they lacked the uniformity and quality that consumers had been able to expect from alcoholic beverage producers.[8]

The Illinois Glass Company also confronted the need to bring in younger management to face the changing industry. But its first generation of leadership stubbornly held on to power. Charles's son William E. Levis sought to replace those who he felt were stifling the company's development. He was an experienced leader who had already seen more of the world than most. After attending military school, William Levis enrolled at the University of Illinois, and eventually in Officers' Training School. He was sent to the front lines of France in World War I, where he received the Distinguished Service Cross for his action during the war. He endured mustard gas attacks and months of combat duty, and his cousin Preston Levis said that he suffered from "shell shock" upon his return to Alton.[9]

William Levis's father abruptly retired from the company at the age of 60, but his uncles were less willing to turn over the reigns. They offered to give William the title of president, but only as figurehead, and he refused. Finally, in 1928 his uncles relinquished control and at the age of 30 William Levis became the president in both name and responsibility. He brought other younger leaders into the company, including his cousin Preston, and Harold Boeschenstein, whose father was a newspaper publisher in Edwardsville, Illinois, just south of Alton. The two had been friends for many years.

Under Levis's leadership, the company set its sites on the Owens Bottle Company. While Illinois Glass was larger, Owens owned the patents on most of its automatic bottle machines and collected royalties from any company that used them. To compete, Illinois undercut Owens on bottle prices, and even undermined Owens's monopoly on one of its most profitable accounts, the emerald green glass used by Canada Dry bottlers. Illinois Glass worked hard to develop glass of such a similar color that it could not be distinguished from the Owens bottles. Illinois also continued to buy up other bottle companies.

Without the drive of Michael Owens, the Owens Bottle Company was feeling the heat of competition from Illinois Glass. William Boshart, who had assumed the presidency of Owens Bottle after the death of Edward Lib-

William Levis, who became president of the Illinois Glass Company in 1928, and president of Owens-Illinois in 1930. (Ward M. Canaday Center for Special Collections. Used by permission of Owens-Illinois.)

bey, was anxious to save the company. In November 1928, Boshart received an internal Owens document outlining the advantages of merging with Illinois Glass, primarily because the market could not support two large glass container companies, and executives feared that the "aggressive policy of expansion" by Illinois threatened Owens's survival.[10] A merger would also allow for elimination of duplicate facilities, ensure control of needed raw materials, merge research and development of new product lines, and broaden sales into new territories. The suggested price was $19 million.

Negotiations on the merger were complex because of the expansive businesses of both. Also of concern was who would be in charge of the new company. Boshart had been president of the Owens Bottle Company since 1925. But the Illinois Glass Company had younger, more aggressive managers who had already shown their desire to control the glass container industry. The agreement finally announced in March 1929 indicated that the company was to be named Owens-Illinois. Furthermore, it said that William Boshart would have the title of president of the company, and that William Levis would become vice president and general manager, a member of the board of directors, and a member of the executive committee. Harold Boeschenstein, Levis's longtime friend, would come with Levis as general sales manager. The company's offices would be located in Toledo, and "a number of the Illinois Glass Company's officials and employees will be merged with the local office force."[11] Stockholders in the Owens Bottle Company assembled on April 17, 1929, in the Toledo Chamber of Commerce auditorium to approve the merger. While most in Toledo assumed that Owens was taking over Illinois, in actuality it was the other way around. This would soon be discovered as the Illinois management team quickly moved to exert its influence.

While Boshart had the title of president, Levis considered him little more than a figurehead. Almost immediately, Boshart and Levis began to quarrel over the speed at which critical decisions were being made. Eventually, the situation came to a head, with Levis threatening to quit unless Boshart resigned.[12] The directors instead convinced Boshart to accept a deal of six months salary in exchange for his resignation. Reports in the local media said Boshart resigned for health reasons. William Levis was named the new president of the company.

The company's annual report to stockholders of 1929 included none of this difficult behind-the-scenes maneuvering. Instead, the report focused on the expansion of product lines with the merger.[13] Owens-Illinois was now the country's largest manufacturer of milk bottles, domestic fruit jars, and wide-mouth containers. The Chicago Heights Bottle Company that was part of the deal produced handblown, high-quality perfume bottles and other specialty products. Plants were located in Alton, Illinois; Bridgeton, New Jersey; Charleston, West Virginia; Clarksburg, West Virginia; Evansville, Indiana; Gas City, Indiana; Glassboro, New Jersey; Huntington, West Virginia; Loogootee, Indiana; Minolta, New Jersey; Newark, Ohio; Okmulgee, Oklahoma; Streator, Illinois; and Toledo. "The management of the

Company is exerting itself to the utmost to reflect the advantages of the Company's improved position in the higher quality of its products, better service to its customers and a greater return to its shareholders," the report stated.[14] Even the company's logo was changed to reflect the merger. An "O" was incorporated around the "I" of the Illinois Glass Company logo, with both overlaying the diamond shape that had been a part of the Illinois Glass logo. The company's product line was given the name "Onized."

A New Company—and New Problems

Among the first issues to confront the company was whether or not Owens-Illinois should try to buy back all of the licenses it had sold to other glass container manufacturers. Levis backed the idea, which took about five years. The company added about $100 million worth of new companies to the Owens-Illinois portfolio through these purchases. Another issue was how many people from Alton would be moved to the Toledo headquarters. Quickly it became clear that most of the managers of the new company were from Illinois, which created some resentment among the Owens people.

Even Michael Owens haunted the merger. The contract signed with Owens in 1922 was to pay him a royalty of one cent per gross of everything produced by the company until 1939, which would have cost the company $1 million per year. That contract was a part of Owens's estate. William Levis wanted to get out of the contract, and proceeded to fight with the Owens estate, claiming that the contract was invalid. Levis stopped all payments. This caused additional strains between the company's new managers and those who still fondly remembered Owens. Finally, in 1931, Owens-Illinois settled with the estate for $500,000.[15]

Six months after the new company was formed, the country was hit by the stock market crash of October 1929, and the start of the worst economic depression of modern times. The company's first merged annual report noted that conditions in the industry were such that "it was impossible to operate all plants in full. It was necessary to concentrate operations in those factories which were best suited to giving our customers service, and where operating costs were lowest."[16] The plants at Loogootee, Indiana; Newark, Ohio; and Minolta, New Jersey were closed, and the plant at Toledo was dismantled. By 1930, the full effects of the Depression were felt. "Market prices for glass containers have continued to decline. Today the price levels in many of the company's lines are the lowest the industry has known for twenty years. . . .

There is a general recognition in the trade of the unprofitable level of prices and, while the movement is still downward, in some of the lines we are of the opinion that it has reached the bottom," the annual report noted.[17] By 1931, the plants at Clarksburg, West Virginia; Okmulgee, Oklahoma; Brackenridge, Pennsylvania; and Evansville, Indiana joined the list of those that were closed. Earnings for 1932 continued to decline, down to $2 million from $2.74 million the year before. The Depression also impacted workers in the glass industry. For example, the average weekly salary for members of the AFGWU declined from $30.77 in 1932 to just $17.90 in 1933.[18]

Fortunately, good news came in 1933 with the repeal of Prohibition. Even before President Franklin D. Roosevelt took office, Congress introduced a resolution to repeal the Eighteenth Amendment. States began to meet in conventions one by one to vote on repeal, with Utah pushing the repeal over the top in December. The great national experiment ended as a miserable failure. One of the motivators of states to repeal Prohibition was the hope that new tax revenues from alcohol sales would boost their cash-strapped treasuries.

Of course repeal could not happen overnight. The Roosevelt administration had to plan how best to implement in an orderly way the flood of alcohol that was sure to hit the market. The repeal of Prohibition meant the beginning of a complex system of federal regulation for the distillery industry. To prepare, Secretary of the Treasury Henry Morgenthau Jr. called for a meeting of bottle producers in 1933. Owens-Illinois was represented by Smith L. Rairdon, sales manager for pharmaceutical bottles. His representation was not as unusual as his title might imply, since the company had been producing bottles for "medicinal whiskey" throughout Prohibition. With Prohibition ending, Rairdon was put in charge of O-I's new liquor bottle division.

The government wanted a way to make sure that the alcoholic beverage industry did not produce tainted products (a problem during Prohibition, which resulted in the deaths of many), that the bottles could not be reused by the many moonshiners who had been making their product for the past decade who would likely try to continue to sell their liquor, and that the new legal liquor industry would pay the required taxes. Rairdon suggested that the industry set standards for sizes and shapes of the bottles so that illegal bottles could be more easily detected.[19] He also suggested that somehow the government had to ensure that the bottles could not be refilled by illegal producers. He hit on putting plant-identifying numbers on each bottle that indicated the month and year each bottle was produced and where. Each bottle had to display the words "Federal Law Forbids Sale or Reuse of this Bottle." All bottles had to be destroyed when empty. The law went into effect in 1935, and many people who had been buying up used bottles and reselling them were put out of business.

In addition to hard liquor, the brewery industry pushed to increase sales by selling beer in attractive bottles with eye-catching labels. Prior to Prohibition, beer had been sold at the local tavern in everything from buckets to large, refillable growlers. As a result, it was not seen as a beverage for the table or to be drunk with meals. Brewers felt that improving the containers would allow a larger market for their product. The trade publication *Brewer and Maltster* featured articles in 1934 and 1935 entitled "Sell Bottled Beer."[20] The articles reprinted advertisements developed by some breweries that focused on increasing the market for beer by touting its health benefits, its sophistication, and its cheapness. The articles stressed that the home market for beer was the largest unexplored market for the product, and the key to expanding this market was to make the product attractive to the housewife. "She must be educated to a more easy use of the word, beer, just as she has been educated to the easy use of the word, cigarette," the articles noted. "The beer bottle and label are equally important. If the bottle is clear and clean and the label attractive, the housewife will enjoy placing the bottles upon a tray for serving in the home."[21]

Such advertising worked. The repeal of Prohibition brought back profits to Owens-Illinois that had been declining from the one-two punch of Prohibition and Depression. The company's 1933 annual report noted that the bottle plants were again operating near capacity, but not enough to open the closed plants. Profits for the year were $6 million, compared to $2 million the year before.[22] The 1934 report noted the reopening of Charleston plant, which produced liquor bottles nearly exclusively. Profits for 1934 were up to $6.5 million.[23]

Diversification of Products

The downturn in the demand for bottles that had been brought on by Prohibition led William Levis to realize that the company needed to diversity its product line. One of the first new products developed was a structural glass block. The blocks were made of two halves sealed together, and could be mortared to build walls just like clay bricks. But because of the air space between the two halves, they had better insulating properties than clay brick,

The production of glass block, ca. 1935. (Ward M. Canaday Center for Special Collections. Used by permission of Owens-Illinois.)

and they also allowed in light. The blocks also complemented the architectural styles of the period by looking sleek and modern. The company began to develop glass blocks in 1931, and by 1932 had a plan to market them.

Because it was difficult to describe in a sales brochure how glass blocks would look, and distributors of traditional clay bricks had no experience with selling such items, the company decided to create various installations for its Insulux Translucent Masonry as a marketing ploy. For maximum national exposure, the company constructed a building made entirely of the product for "A Century of Progress," the world's fair that was to be held in Chicago in 1933.

"A Century of Progress" was organized to celebrate the 100th anniversary of the founding of Chicago. As with most world's fairs, it promoted optimism for the future. The official theme was a celebration of the advancement of mankind during the previous 100 years. Its message of hope and faith in the future was especially needed given that the fair opened during some of the darkest days of the Depression. To stress its modernistic vision, the exposition buildings were all constructed in the modern architectural style. As one of the many guidebooks to the fair noted, "It would be incongruous to house exhibits showing man's progress in the past century in buildings which were reproductions of ancient Greek temples or Roman villas. The architecture should be in tune with the modern age. It should express through its designs the needs of the present, and it should anticipate so far as possible the requirements of the future."[24] It was a perfect setting for a building built entirely of Owens-Illinois's newest product.

Designed by architect Eloy Ruis, the O-I building was lit with electric lights at night, and seemed to glow as if a beacon. The company noted that while built as a temporary structure, it withstood Chicago lakefront weather because of the insulating properties of the glass block. Inside, the building included displays of various O-I products, including a panorama scale model of a glass bottle factory. It also included examples of other O-I products such as vacuum-packed coffee in glass containers, "Presto" canning jars, glass containers of motor oil ("the motorist not only sees what he buys, but he knows what he pays for"), and a brand-new product—furnace air filters made out of finely drawn strands of glass.[25]

The company also used the interior of its headquarters in the Ohio Building in downtown Toledo in 1935 to show off its new product. Private offices were constructed of the glass blocks. The reception room was built exclusively of glass blocks, and its ceiling was built of glass acoustical tile. The tile helped to muffle office noise. Glass panels around the room depicted the history of glass from ancient times to modern.[26] Photos of the offices show an odd mix of sleek, modern glass block walls and traditional wood office furniture.

After the success of the Century of Progress glass block building, the company built a research laboratory in Toledo in 1936 that was completely windowless and constructed exclusively of 80,000 Insulux glass blocks. The building was built on Westwood Avenue near Dorr Street in west Toledo. It had partitions also built of glass block so that light could flood into the center of the building. Because the building had no windows, it was air conditioned—another progressive building design feature. Insulation and air filters used in the building's construction were made of glass fibers then under development by Owens-Illinois. The building was the first permanent structure built of Insulux block, and it would serve as a model for another research laboratory constructed in Newark, Ohio, in 1937.

Glass block buildings became popular for their modern, avant-garde look. The first all-glass block home was built in 1937 in Marblehead, Massachusetts, by Aroline Gove, daughter of Lydia Pinkham, the medicine company heiress. The blocks also were incorporated as building elements in many homes, public structures, and office buildings in place of windows, or to add drama to architectural features such as entryways.

A New Product—and Another New Company

The basic method of producing fibers from glass has been known for centuries.[27] The Venetians, the most skilled of early glassmakers, drew glass into fine strands to decorate their vessels. While the Libbey Glass pavilion at the 1893 World's Columbian Exposition featured a dress woven of glass fibers, Libbey's glass fibers were mostly for show, and had little practical application because the fibers would easily break, and cloth made from them was heavy and difficult to sew. The company discontinued work on developing glass fibers after the fair.

The early method for making glass fibers was described in an article in the *Pittsburgh Gazette* in 1880.[28] Two workers would start by pulling a thread from a single gob of glass by walking across the production room in opposite directions to create a rope of glass. The rope was cut into sections, and then taken to the spinning room where a hot-air blowpipe was set up next to a steam-powered spinning wheel. As the rod was heated, the steam-powered wheel would spin the rope into fine fibers. The faster the wheel turned, the finer the glass thread that was produced. In Europe, Oscar Gossler developed a process for making glass wool from leftover molten glass rather than using glass rods. But this process produced a heavy wool weighing 13 pounds per cubic foot, hence limiting its applications.[29] Other attempted processes used centrifuging—spinning glass fibers off a large, horizontally spinning wheel.

As with the production of glass block, fiberglass experimentation at Owens-Illinois was the product of excess production capacity during Prohibition and the Depression and the need of O-I to diversify its product line. The company's equipment was deteriorating from lack of use, and something had to be done. Producing glass fibers appeared to Levis like a sound investment because fibers of glass had seemingly limitless industrial applications. They were flexible yet strong, lightweight, waterproof, heat resistant, and immune to interactions with other chemicals. But whether or not glass fibers could be produced in large enough quantities and at a cheap enough price to make them commercially viable was unknown. Experimenting with this new product line was risky business for the company.

Fiberglas in production. (Ward M. Canaday Center for Special Collections. Used by permission of Owens Corning.)

To help make the decision about whether fiberglass research was a sound investment, William Levis brought in the consulting firm of Arthur D. Little from Boston. Little, in turn, contracted with two glass experts, Leonard Soubier and Joseph Wright, to experiment producing rock wool, or rough wool made of glass, at the company's Alton factory in 1931. Their process involved a stream of glass from a glass furnace subjected to a stream of air. While the process looked promising, it was still far from commercially viable. But it was successful enough that the company decided to set up a separate research facility at the Evansville plant to continue experimenting with production methods.

The research on glass fiber production was put under the direction of Dr. Games Slayter, who had been a consulting engineer in Detroit, where he worked in developing insulation. One of his products was a spongy glass foam block, which is what attracted the attention of Owens-Illinois. He was assigned a full-time assistant, John H. Thomas. Both were transferred to the Evansville research facility.

The first commercial application that O-I sought to develop was glass wool for air filters. The market for air filters was being driven by the forced air furnace industry, which was trying to market itself as an alternative to steam heat. But forced air furnaces were dirty because they circulated a home's dust around as they blew in hot air. The solution was a filter, and the one Slayter and Thomas invented in 1932 could be made cheaply. It was made of a series of bonded glass fiber mats. The product was named the Dust-Stop filter, and its advertisements said it was made of a new product called Fiberglas.[30] It was advertised as fire resistant and effective at removing dust and allergens from the air. A few years later, the company replaced the expensive metal covers on the filters with the blanks left over when bottle caps were punched out of a sheet of metal, which was not only cheaper, but also a more attractive cover.

While the Dust-Stop filters were successful in doing what they were intended for, there was not sufficient demand to keep Fiberglas profitable. New uses were needed. Research on Fiberglas was moved from Evansville to Columbus, Ohio, in 1932. It was here that the first real breakthroughs occurred in developing a new and potentially more profitable product, insulating glass wool produced from continuous fine glass fibers. But these breakthroughs did not happen easily. The most difficult process to perfect was the spray guns used to blow the fibers so that the fibers would remain consistently fine. The work on the process was hot and dangerous, as the guns would clog and required operators to pull out the fibers by hand.[31] Slayter remembered a process that he had developed to apply the name of dairies to milk bottles that used an acetylene flame to force glass droplets into fibers, and began experimenting with this method. He assumed a bigger acetylene gun might work to produce more glass fibers, so the research team developed a much larger one. Unfortunately, during a demonstration of the process, the gun became coated in glass and exploded, nearly destroying the Columbus facility.[32]

After the disaster at Columbus, the research project was moved due east to O-I's Newark plant. At one time, the Newark plant had been one of the largest producers of glass bottles in the world, employing 1,700 workers. But because of Prohibition, the factory was shut down in 1930. Those who had shuttered the plant had done so quickly, without cooling the furnaces—they were simply abandoned.[33] The result was that the furnaces ruptured, spilling molten glass. The plant was a disaster. But it was available for experimenting, and unlike Columbus, any explosions would not disrupt other production.

Slayter and Thomas gave up on acetylene and returned to steam. This idea came to them while watching Dale Kleist, an assistant to Thomas, try to weld two glass block halves together using a stream of glass. But all Kleist could produce using a metalayer gun and steam were fine fibers. Slayter and Thoms realized that if steam could be applied to a fine stream of molten glass in a consistent manner, smooth continuous fibers could be produced that could be used in commercially viable insulation.

But there were still major issues to overcome. One was how to protect the stream of glass as it came out of the furnace. The only product that seemed to work was platinum, which was expensive, and did not last long. Using alloys of platinum helped extend its life and lessen the cost. But as the glass hit the platinum, it cooled, producing chunks of glass in the wool, which were unacceptable. Slayter suggested heating the platinum using an electrical current so that the glass could flow at a constant temperature in a steady stream. Each orifice from which the glass emerged could be maintained at the same temperature, so that all fibers would be consistent. With these improvements, Owens-Illinois had a commercially sellable product.

In a radio address in 1936, Faustin J. Solon, vice president and general sales manager for O-I, described the wonder of glass fibers to listeners. "Conceive, if you will, a single strand of glass, over thirty-one million feet long, practically six thousand miles, from a pound of glass. Think of it—one pound of glass, no more than is in a ketchup bottle, forming a thread that would reach from New York City to Los Angeles and back again," Solon boasted.[34]

While there had been some victories in developing glass fiber technology, they had come at a huge cost. And before the products could be sold, these technological problems had to be overcome to allow mass production. William Levis was friends with Amory Houghton, head of Corning Glass in New York. Corning had also been experimenting with glass fiber production, with limited success. In 1935, Levis and Houghton agreed to pool their research efforts. This move brought many high-quality glass researchers from Corning into the efforts to produce glass fibers. A deal struck in October of that year gave Corning rights to all O-I glass fiber patents, and O-I received the same from Corning. Engineers from Corning were sent to Newark to learn what Slayter, Thomas, and Kleist were doing.

Researchers at both companies met frequently to share ideas. A new research facility was constructed in Newark in 1937 for further development of the Fiberglas product after another fire burned the original building. The research center was featured in an Owens-Corning marketing campaign titled "Alice in Wonderland." It depicted an attractive young woman named

Alice, dressed in childlike clothing, who was photographed watching the various stages of Fiberglas production and marveling at the miracle of the product.[35] The dedication of the Newark laboratory was attended by scientists, educators, and members of the press from around the country, and hosted by William Levis.

Now that the technology was moving forward, the companies had to begin selling Fiberglas. The only two products that had been successfully marketed were the Dust-Stop filter and home insulation. Home insulation was sold exclusively by the United States Gypsum Company under the trade name "Red Top Insulating Wool." Like the Dust-Stop air filters, Red Top insulation was advertised for its many advantages: it was lightweight, fireproof, permanent, and economical. "Fiberglas keeps people warmer in the winter and cooler in the summer, and at the same time, saves them heating and air conditioning costs," advertisements for the product noted.[36] U.S. Gypsum's exclusive sales agreement for Red Top meant that Owens-Illinois did not gain much profit from the sale of the product. In hopes of increasing profits, Owens-Illinois also experimented with cloth woven from Fiberglas, but it was devilish to perfect, especially methods for weaving the cloth on conventional fabric looms, which would break the fibers. Other experiments included developing insulation for electrical wires, and using Fiberglas in other composites to strengthen them.

Seven years of investment in Fiberglas had not contributed to the company's bottom line. No products had been developed that could carry the cost of further research and development. Fiberglas remained an expensive scientific experiment that was not the principal interest of either of its sponsoring corporations—Owens-Illinois or Corning Glass. In the spring of 1938, an analysis showed that both companies had lost over $500,000 in the previous year on the development of Fiberglas.[37] While most recognized the enormous potential for developing new products, dividing responsibilities for this development between two companies diminished the chance that either would produce sufficient volume to cover costs. Many felt that it was necessary to turn the research loose to let a new company that produced only Fiberglas sink or swim on its own.

In addition to the bottom-line reason for spinning off Fiberglas as a new company, there was also a personal one. Memos drafted by Harold Boeschenstein to William Levis early in 1938 hinted at a power struggle within Owens-Illinois between the two men. Discord was impacting the management of the company. "I am certain that this situation is rapidly undermining, and will soon destroy, those characteristics of faith in leadership loyalty and team work which have made ours an outstanding organization and a successful business and which have given much to each of us," Boeschenstein wrote in a memo drafted by hand from a hotel room in San Francisco in March 1938.[38] "I believe we have at heart a consciousness of our obligation to the other workers and investors in the business. They have a right to expect from us unselfish leadership and an honest example. Even in the absence of this sense of obligation, a wholly selfish consideration of the fact that each of us, and our families, are dependent upon this business for large part of what he have, should prompt us to recognize that our leadership is on trial and the future of this company is at stake," Boeschenstein darkly warned.[39] It is unclear if Boeschenstein ever sent this memo, or others that he drafted that year. One from April 1938 continued to lay out the problems for Owens-Illinois, including rapid diversification, an aging workforce that lacked enthusiasm, and the depressed economy. In this memo, Boeschenstein enumerated the advantages of a new Owens-Corning company: economical production, coordinated development of products, and elimination of conflicting marketing policies. Boeschenstein also expressed his own personal frustration of working with William Levis. "What is my future in this business? Have made it my job to complement and supplement—not try to take the ball and run as my own independent judgment has prompted. Now, find myself in an ambiguous and what may become an untenable position—and not clear, as I once was, as to your thinking or your objectives. . . . I don't know how to play the game with you."[40]

By May, it appeared that the idea to create a new company had been decided. A memo drafted by Boeschenstein stated his demand that he become president and general manager of the new company.[41] In the rest of the draft to Levis, Boeschenstein laid out his compensation and stock option requirements for assuming the position.

A report prepared jointly by Carl Megowen, comptroller at O-I, and A. C. Freligh, accountant at Corning, in August 1938 supported the creation of a new company separate from the two other companies. "We have investigated the present state of fibre glass development and have also estimated long term potential sales and near future profits on the assumption that the divisions of both companies might be consolidated into a single independent operation. On the basis of this investigation and the estimates, we recommend the immediate organization of a corporation to own and operate the two divisions, which corporation shall be independent and under separate management."[42] The report listed anticipated new potential markets for Fiberglas: textiles, pipe insulation, insulating wall boards, railway insulation, acoustical tile, and shingles.

Harold Boeschenstein and Games Slayter, who was called the "Father of Fiberglas," ca. 1940. (Ward M. Canaday Center for Special Collections. Used by permission of Owens Corning.)

On November 1, 1938, Harold Boeschenstein announced the new company on letterhead that read the Owens-Corning Fiberglas Corporation, the creation of which had been finalized the day before. "Owens-Corning Fiberglas Corporation is the outgrowth of seven years of experience in research, development and successful commercial demonstration in many applications," Boeschenstein wrote. "Through coordination and expansion of research and facilities, we anticipate further development of Fiberglas in various forms, its broader application to additional industrial and domestic uses, and economies in operations."[43] The letter indicated that management, research, and operating personnel had been selected from the two companies to form the new company, with the total number of employees around 600. It was signed "Harold Boeschenstein, President." Other leaders included Amory Houghton as chairman of the board and Games Slayter as vice president. William Levis was also a member of board of directors. The company leased offices in the building in downtown Toledo that housed Owens-Illinois.

In addition to the official announcement of the company, Boeschenstein also issued a statement called "The Four Basics" that was to guide the company's product development.

> Either our materials must do a job that no other material can do effectively. . . .
> Or for the same cost they must perform better than competitive materials. . . .
> Or at a lower cost they must do as good a job as alternative materials. . . .
> Or their unique characteristics must enable the manufacturer using Fiberglas to make corollary savings not possible with some other materials.
> This statement will continue to be our challenging pledge for the future.[44]

Owens-Illinois Continues Its Expansion

Continuing his efforts to diversify Owens-Illinois's product line, William Levis set out to purchase other companies that complemented the company's product line. In 1931, Owens-Illinois purchased the Illinois Pacific Glass Corporation, which had been a part of Illinois Glass, thus opening up West Coast markets to O-I. William Levis's cousin Preston was put in charge of the company, and would remain in that position until 1938, when he returned to Toledo to take the position previously held by Harold Boeschenstein as vice president and general manager of Owens-Illinois. Illinois Pacific was profitable for O-I, particularly in the alcoholic beverage bottles market after the repeal of Prohibition.

One of the most surprising expansions was the purchase in 1935 of the glass company that started it all in Toledo—the Libbey Glass Company. Owens-Illinois had been the exclusive seller of some lines of Libbey products for a year, but was interested in gaining control over production as well. Of particular interest to O-I was Libbey's thin machine-blown tumblers and stemware, which were strong yet lightweight and manufactured using the Westlake machine. The products made using the Westlake machine were called "Libbey Safedge" because of a thicker edge that reduced chips, a process invented by A. C. Parker of Libbey in 1923. Libbey had been making Safedge glassware not only for drinking glasses, but also for food packaging, which put the company in competition with O-I. The purchase of Libbey for $5 million gave O-I the largest thin-blown glass plant in the world.[45]

The purchase of Libbey Glass rescued Libbey from difficult financial straits brought about by the disastrous investment in 1933 in a product line called the "New Era in Glass."[46] It was a high-end tableware line designed at considerable expense by A. Douglas Nash, who had operated a glass factory in Corona, New York, for Louis Comfort Tiffany. In 1928, Nash took over the Tiffany factory, and it failed within two years. Nevertheless, Libbey's leader believed Nash would bring Tiffany's success and manufacturing secrets to Libbey. Nash was hired by Libbey in 1931, and he began development of the new line for Libbey with little oversight or cost controls.

Design work was expensive, and production was even more expensive—the company had to borrow between $300,000 and $400,000. The product line challenged the abilities of the Libbey craftsmen. The pieces were intended for formal table services, so each set consisted of at least five different pieces. Some pieces were custom made for individual buyers. It was introduced to the public through a flashy marketing campaign in high-end magazines like The *New Yorker*, *Vogue*, and *Vanity Fair* in 1933, at the depths of the Depression.

The product line was poorly matched to the economy of the country at the time, and it tanked. Libbey Glass took a huge financial hit for the Nash glassware. Owens-Illinois made its offer to buy the company as Libbey struggled to survive the debacle, and Owens-Illinois probably prevented Libbey's demise. Ironically, because the Nash line was of such quality and made in such small quantity, today it is highly sought after by collectors.

Owens-Illinois now controlled all of Libbey's product lines, including its extensive commercial glassware line sold to restaurants, hotels, and bars (now in demand with the end of Prohibition). The acquisition also brought back together the two original companies of Mr. Libbey and Mr. Owens a decade after the death of Libbey. But some members of the AFGWU, especially those who were still producing glassware by hand, were apprehensive about the purchase of Libbey by the company that had automated the glass bottle industry.[47]

Owens-Illinois's 1935 annual report justified the purchase to its stockholders. "While the paste-mold manufacturing process is distinctly different from your Company's iron-mold process, and the business in all its phases is non-competitive with your Company, it is felt that this is one of the divisions of the glass industry that gives promise of substantial future growth. The Libbey Glass Manufacturing Company has developed a manufacturing art and technique in its field which will be valuable in many of your Company's newer developments and in manufacturing of certain styles of lighter weight containers for which there is demand."[48] Profits were helped in 1937 when Libbey won exclusive rights to use the characters from Walt Disney's enormously popular movie *Snow White and the Seven Dwarfs*, which was released that year. Libbey sold tumblers decorated with the characters to food companies, which in turn filled them with products like jelly. Once empty, they could be reused as drinking glasses.

Another Depression-era expansion came in 1936 when Owens-Illinois purchased the Tin Decorating Company, the Enterprise Can Company, and the St. Louis Can Company. The purchase of these companies expanded the O-I product line beyond glassware as containers. The companies were merged into a new company—the Owens-Illinois Can Company. By the close of 1936, Owens-Illinois employed more people than it had at any time in its history—over 18,000.[49]

In addition to acquisitions, Owens-Illinois expanded the product line of its bottle plants. In the late 1930s, O-I became focused on making lighter bottles. Through extensive research, it developed a new way to create both lighter and stronger bottles. To reassure consumers that lighter did not mean more easily breakable, the product line was called Duraglas. Duraglas was made through improvements to the bottle machine that allowed more careful control over production processes. Duraglas bottles had a strong lip that was resistant to chipping, so they could be used several times by bottlers. They could also be printed with color printing that would not deteriorate after repeated washing. In 1936, O-I introduced a new Duraglas beer bottle called the "stubby," which was a no-deposit, nonrefundable bottle that revolutionized how beer was sold in bottles.[50]

Labor in the Glass Container Industry

While it can be argued that the glass container workers were impacted much more directly by automation than flat glass workers, there was less labor unrest among the container workers in the Depression decade. Policies of the federal government that encouraged collective bargaining led to a major strike in 1936 and 1937 in the flat glass industry that impacted the workers at Libbey-Owens-Ford. But there was no such disruption at Owens-Illinois.

When the Owens Bottle Machine was introduced in 1904, many skilled glassworkers were faced with losing their high-paying jobs. And many did. But through expansions of the number of shifts and the elimination of the apprenticeship system, the number of workers in bottle production remained fairly steady as the workweek decreased in hours.[51] Owens-Illinois fought collective bargaining before the Wagner Act made it open up to unionization. But once accepted, collective bargaining actually helped companies like Owens-Illinois in several ways. The equalization of wages across companies helped to stabilize profits and kept some companies from undercutting the competition. In 1934, the GBBA even stated as such in its constitution: "The object of this association is to thoroughly unite all glass workers and other engaged in the industry for their mutual benefit and protection and to regulate and maintain a uniform price list throughout the trade."[52]

A culture of cooperation between labor and management in determining wages helped to settle disputes quickly.[53] "The union president has full authority to decide all such matters referred to him and his decision is binding," a report issued by the U.S. Department of Labor noted in 1936.[54] The stability of the industry was helped by the fact that between 1891 and 1936, only five men held the office of president of the GBBA. In addition, some companies sought to ensure labor peace by hiring those who were active in the union to represent the company in labor negotiations. For example, the labor relations expert hired by Owens-Illinois had been president of the Flint Glass Workers Union for 13 years, and having seen the negotiations table from the other side, was more willing to work to compromise. The industry also used the GBBA to lobby for the repeal of Prohibition, and to eliminate bottle deposit regulations that paved the way for the disposal "stubby" beer bottle. The union also used the tradition of the "summer stop rule," where glass plants shut down between July 1 and September 1 due to heat, to demand for paid vacations, an unusual benefit for workers of the time.[55]

What a Difference a Decade Makes

The 1930s ended with Owens-Illinois as a drastically different company than it had been when the decade began. The merger between the Owens Bottle Company and Illinois Glass Company had been difficult, but successful. The leadership of William Levis had moved the company to the top of the glass container industry. The end of Prohibition meant the return of one of the company's largest product markets, and new acquisitions and new products had diversified O-I's product line so that it could, its leaders hoped, avoid the impact of any future such events. A new spin-off company was developing an entirely new glass-based industry. In 1939, the future for the company, and the country, seemed without limits.

SIX

Building the World of Tomorrow

> GLASS and CIVILIZATION! The two march hand in hand down the corridors of time. Age after age the miracle of glass unfolds with the fabrication of this versatile material in to an endless variety of new forms that weave themselves ever more widely and firmly into the pattern of life.
>
> —*The Miracle of Glass: Its Glorious Past, Its Thrilling Present, Its Miraculous Future as Presented at the Glass Center*, New York World's Fair, 1939

"I Have Seen the Future." Buttons sporting this marketing slogan became some of the most sought-after souvenirs of the 1939 New York World's Fair. The buttons were handed to fair visitors as they exited "Futurama," the popular attraction sponsored by General Motors. Futurama was the vision of theatrical designer turned urban planner Norman Bel Geddes of what the United States might look like in the year 1960. A diorama of cities, farms, and suburbs made up of 500,000 miniature houses and 50,000 to-scale automobiles, Futurama promised a well-organized future America of peace and prosperity where everyone was linked by an elaborate system of superhighways.[1] While the actual future did not live up to the ideal of Bel Geddes or other exhibitors who displayed similar utopian dreams, the 1939 World's Fair brought together a unique confluence of people, ideas, and marketing that would have a tremendous influence on postwar America, including Toledo.

The theme of the fair was "Building the World of Tomorrow." The focus on the future was hardly surprising, given the state of the country and the world. A decade of economic depression had shaken the populace to the core. In America, this disillusionment was particularly pronounced. For the country to survive, political leaders re-created the government through a series of New Deal agencies that brought the federal government into aspects of life never before imagined. Some felt these rescue programs did not go far enough, and some social activists and labor leaders questioned whether democracy and capitalism should be saved at all. On the world stage, autocratic leaders used depression and discontent to consolidate power, with military plans to grab more. "Building the World of Tomorrow" was a far more optimistic goal to most fair visitors than dealing with the reality of the world of the day.

For fair planners like Grover Whalen and Robert Moses, "Building the World of Tomorrow" meant imagining a world where science and technology would be harnessed for the good of all.[2] But not pure science and technology—rather, science and technology applied to consumer goods produced by America's corporations to improve lives individually and society collectively. In this new future, corporations would not just produce goods, but they would be important social institutions, where workers prospered through their role in the corporate state.

Toledo Goes to the Fair

Toledo's glass corporations were well represented at the 1939 New York World's Fair. This was, of course, not the first time that the glass companies had created large displays at such gatherings to promote their products. Similar displays had been developed for the 1893, 1904, and 1933 fairs. At the 1939 fair, two of the city's glass manufacturers, Owens-Illinois (including its wholly owned subsidiary, Libbey Glass) and Owens-Corning Fiberglas, joined with Pittsburgh Plate Glass to create the Glass Center. As it had been at the 1933 fair, the building that housed the glass companies' displays was designed to showcase their products. The round building was designed by the firm Shreve, Lamb, and Harmon, the same architectural firm that had designed the Empire State Building. It was adorned with a tall glass tower and constructed of windows, mirrors, fiberglass, and insulated glass block. The building had 25,000 square feet of display space surrounded by 7,000 square feet of glass block, 7,000 square feet of plate glass, and 6,000 feet of structural glass.[3]

Inside, the Glass Center exhibits showed the importance of glass

The Owens-Illinois display inside the Glass Center at the New York World's Fair, 1939. (Ward M. Canaday Center for Special Collections. Used by permission of Owens-Illinois.)

throughout history, with a special emphasis on its imagined importance to the future. The rotunda featured a glass furnace where artisans produced beautiful handmade objects one at a time, just like the Libbey Glass pavilion at the 1893 World's Columbian Exposition. But the exhibit went beyond blown glass to show pressed, rolled, and drawn glass.

In the book *The Miracle of Glass*, available to visitors of the Glass Center as a keepsake to take home with them, those who missed the opportunity to visit the fair were treated to breathless descriptions of the beauty of the Glass Center. "After seeing glass, familiarly known as a hard, resisting substance, woven into soft and pliable fabrics, we are mentally prepared to expect practically anything of this versatile material. With skepticism reduced to the vanishing point, we mount a sparkling glass stairway and emerge into the open air onto a glass ramp, from which we can now view with equanimity the fantastic symbolic decorations on the patio walls. Here imaginative artists inspired by past and present wonders of glass fabrication, have depicted for us in glowing color and futuristic line a suggestion of the limitless possibilities of glass in meeting the now and faintly envisioned requirements of the World of Tomorrow."[4]

Owens-Illinois described its exhibit in equally breathless language. "As you enter the rotunda of 'Glass Center,' with its mirrored glass ceiling, you view an old time glass operation—furnace—molds—tools—men making tumblers, vases, and other objects in much the same fashion as for thousands of years prior to the advent of the Owens automatic machine. Just before you leave, in a section adjacent to walls of very modern Insulux Glass Block, you will see glass marbles melted down and drawn into fine filaments from which thread is made and cloth is woven," the company's annual report of that year boasted.[5] Among the products displayed were bottles decorated with an applied color label that "[helps] producers perpetuate their identity, and gives added assurance to consumers."[6] Included were glass food containers, and even cigars sold in glass containers. The new, lightweight bottle that Owens-Illinois had developed was also featured. The Insulux glass block display noted that in only four years since its inception, glass block had been installed in some 50,000 locations covering every state in the nation. "Authoritative comment is to the effect that no other new building material had ever received so instantaneous and complete acceptance as has Insulux Glass Block," the company's fair publication noted.[7]

Owens-Illinois also displayed products from its newly acquired Libbey division. The Libbey display featured a machine showing how Safedge glassware was formed, and included a display of a hammer repeatedly striking a Safedge tumbler without chipping the glass. The Safedge line sold to commercial outlets was exhibited, as well as a new line of Libbey retail glassware for the home.

For Owens-Corning Fiberglas, the New York World's Fair was the first chance to showcase the company that had been formed only the year before. One display showed men dressed like scientists in white laboratory coats spinning glass onto spools and weaving the fibers into fabrics on looms. Fairgoers could purchase pieces of fabric woven of the new product. Also displayed were the company's only two marketable products of the time—Dust-Stop furnace air filters and Red Top insulation. The company described its insulation as "Wool from Mechanical Sheep" and the machinery that produced it as "Steel Silkworms."

Other displays at the Glass Center showed glass in almost every room of the home of the future—glass block in bathrooms, plate glass in living room windows, glass fibers in attics as insulation, living rooms with furniture that had arms and legs made of glass (including a piano supported by a plate glass frame and legs), and walls adorned by Fiberglas curtains. All were intended to show that while glass had been around for thousands of years, it was not a product of the past, but one of the future.

Destroying the World of the Day

While the New York World's Fair was focused on "Building the World of Tomorrow," others were focused on destroying the world of the day. In 1938, Adolf Hitler claimed Austria for his growing German empire, and was given the Sudetenland by the Munich Agreement. On September 1, 1939, he invaded Poland, and within three months had taken Denmark, Norway, Belgium, Luxembourg, the Netherlands, and France, followed by Yugoslavia and Greece. The United States was trying hard to keep out of Europe's conflict, but President Franklin D. Roosevelt realized that the only country standing in the way of Hitler's total domination of the continent was Great Britain, which Hitler would soon begin assaulting from the air. It was no longer a question of if the United States would be drawn into the war, but when.

For the glass companies of Toledo, the years before World War II were ones of relative prosperity. The Depression decade had been difficult, but all of the companies had expanded during the period, and a new one—Owens-Corning Fiberglas—had been created. Libbey-Owens-Ford had found its major markets to be the automobile industry, housing, and commercial construction. Owens-Illinois had developed Duraglas containers, and was actively marketing them to consumers, especially women. The company even established something called the Home Makers' Guild of America, a group of "housewives" and home economists drawn from a cross section of geography locations, socioeconomic classes, and ages who voluntarily tested O-I products and offered reviews on packaging. "The pulse of Mrs.

The Libbey-Owens-Ford plant in Rossford, ca. 1941. (Ward M. Canaday Center for Special Collections.)

Homemaker has been and is constantly being taken. Her response to any suggested change or her request for a change has guided the manufacturer in his designing of glass containers for her use," a company publication of 1941 noted.[8] But ominously, the company also advised homemakers to create an "emergency shelf" of preserved food products (preserved, of course, in Owens-Illinois glass containers).

In August 1941, Libbey Glass was featured in a five-day series of articles by noted nationally syndicated journalist Ernie Pyle. During a visit to the Libbey factory, Pyle watched as the glassmakers crafted items, and Pyle himself had a chance to try glassblowing. The columns were published in Scripps-Howard newspapers around the country, and after Pyle's death in 1945, they were compiled by Libbey as a book titled *Ernie Pyle on Glass.*[9]

Owens-Corning Fiberglas was devoting its efforts in the prewar period to perfecting mass production and improving its products and marketing. It had developed a new product consisting of insulation applied to semi-rigid boards that was used as refrigerator insulation. It was also producing acoustical tiles, pipe insulation, and construction products. Significant efforts were also going into its new textile line that used continuous fibers woven into fabrics for products such as curtains, boat and automobile covers, and reinforced plastics.

But the companies' leadership could not deny that world events were spinning out of control, and soon would likely draw the United States into war. This reality was especially impactful on Libbey-Owens-Ford. For the second time, John Biggers, president of L-O-F, was called into national service, this time to help oversee the country's preparedness for war. Biggers had first served the Roosevelt administration in 1937 and 1938 as administrator of the Unemployment Census, where he organized the first systematic effort to collect statistics on the unemployment rate of the nation. He was a "dollar a year" employee, a unique position granted by President Roosevelt to businessmen voluntarily willing to serve their country. After completing this job and returning to Toledo, Biggers was contacted by Secretary of the Treasury Henry Morganthau and asked to help oversee aircraft production in France and Britain to ensure that if United States needed planes in case of war, adequate numbers could be produced.[10] At that time, the United States purchased most of its planes from abroad due to its small domestic aerospace industry. But because the Justice Department was in the middle of an investigation of Libbey-Owens-Ford for unfair trade practices, Biggers declined the position.

In 1940, President Roosevelt appointed William S. Knudson, chairman of General Motors Corporation, as chairman of the National Defense Advisory Committee, a civilian committee with responsibilities over defense production. Knudson was first brought to the attention of Roosevelt by Biggers, who recommended him highly. The committee was to study what was required for the country's defense in case of war and aid the military in the acquisition and production of war equipment. Because of his respect for Knudson, Biggers agreed to come to Washington again as another "dollar a year" executive, and worked as Knudson's chief assistant. But the National Defense Advisory Committee was soon replaced by the Office of Production Management in late 1940 when labor criticized the Roosevelt administration for the lack of input from the labor sector in the committee's decisions. Knudson was put in charge of the new office, and Biggers was named director of production. Later, other Toledo glass executives would serve on the War Production Board, which advised the Office of Production Management, including William Levis and Harold Boeschenstein. Boeschenstein would become vice chairman of operations for the board in 1944, the number three position.[11]

The Office of Production Management was charged with gearing up

Women at work on airplane windshields for the war at Libbey-Owens-Ford, ca. 1942. (Ward M. Canaday Center for Special Collections.)

domestic military production, and was given authority to grant contracts. One of its most daunting tasks was to increase the production of airplanes within the United States. In 1940, President Roosevelt requested 25,000 planes, 10 times more than had ever been produced by U.S. manufacturers. The Office of Production Management helped domestic manufacturers to meet that goal, and also helped to establish army tank production.

Biggers used his role behind the scenes to help Toledo companies secure defense contracts. Champion Spark Plug was given a contract to produce aviation spark plugs, which allowed the company to expand into a new factory. But perhaps Biggers's most important advocacy for Toledo businesses was the assistance he provided to the Willys-Overland Company.[12] In 1940, the government put out a call to automobile manufactures to build a prototype for a lightweight but sturdy all-purpose military vehicle. The company that produced the best prototype was Bantam, located in Pennsylvania. But the government felt this small manufacturer could not produce 75,000 of the vehicles, as it demanded, so alternative producers were sought. Also submitting prototypes were the Ford Motor Company and Willys-Overland. Ward M. Canaday, then president of Willys-Overland and a neighbor of John Biggers, asked Biggers how his company could get the contract since Toledo desperately needed the business. Biggers advised Canaday to drop the price, and Willys-Overland came back with a proposal that was $525,000 less than Ford's proposal. But the military gave the contact to Ford.

Biggers spoke with William Knudson and convinced him that Ford was needed for other war products, and that Willys-Overland was the best company to produce the new general purpose vehicle, which came to be known by the name Jeep. Knudson's Office of Production Management overruled the military and granted the contract to Willys-Overland, although Ford also manufactured some Jeeps.

In August 1941, with war with Germany looking more and more certain and Britain struggling to withstand the German onslaught, President Roosevelt sent John Biggers to England to work with W. Averill Harriman, the president's special representative to that country. "At this time it is of increasing importance that the production of war material in America and Britain be synchronized and developed in balance with overall needs," Roosevelt stated in a letter to Biggers advising him of his new assignment.[13] Biggers was to study production in Britain and report back to E. R. Stettinius, the administrator of the Lend-Lease program. The Lend-Lease program was a special effort by the Roosevelt administration to support England through lending and leasing military equipment without actually entering the war on the side of Great Britain. As Biggers later recalled, "My prime assignments were to persuade the British Government and British industry to use machine tools more effectively."[14]

Biggers was given an office in the American Embassy, and even assigned a bomb shelter near St. Paul's Cathedral in London to protect against German bombs that frequently rained down on the city. Biggers also had a chance to meet on several occasions with the respected British prime minister, Winston Churchill. As Biggers later recalled, "Quite apart from the luncheon at his home, his country home, Checkers, and the luncheon at 10 Downing Street and the subsequent dinner at 12 Downing Street at which he took such an important part, I had a meeting with him alone in his office at 10 Downing Street the day before I left England and had a chance to review a number of subjects with him and, of course, learned a good deal from his wide experience in public life and his subsequent handling of some of the problems with which we were then struggling."[15]

In the fall of 1941, Biggers asked President Roosevelt to be relieved of his duties with the Office of Production Management because his wife was ill and his company needed him in Toledo. Few of the "dollar a year" executives were being released from their government contracts at the time. President Roosevelt refused to allow Biggers to resign, but did allow him to take an extended leave of absence because of his wife's condition. "I thoroughly understand your desire and personal obligation to return at this time to your business. After seventeen months of working for the Government, this is wholly proper. But, while you say that if you are needed in Government again you will be ready, I want to put it just a little differently. I am very certain that the Government will need further service from you. Therefore, I feel that I am, in effect, lending you back to your own business—a leave of absence, subject to recall," President Roosevelt wrote to Biggers on October 30, 1941.[16] In just a little over a month, the United States would be swept into the war. If Biggers had not been able to return to the country in October, there was little chance he would have been let go after war was declared. He returned to Toledo on November 1, 1941.

In addition to the pressing personal reasons, Biggers also was needed back in Toledo to gear up Libbey-Owens-Ford for war production. "Good progress had been made by my associates in this direction, but with Pearl Harbor and our involvement in war in December 1941, the whole picture radically changed and we had to largely shut off civilian production and try to transform a glass company, not normally suited to the production of things needed in military conflict, into making a wide variety of things from bullet resistant glass to aircraft windshields and to plexiglass for use in aircraft," Biggers recalled.[17]

Toledo's Glass Companies Go to War

As Biggers noted, glass was not generally thought to be a wartime defense-related product. But if Toledo's glass companies were to survive with the dramatic cutback in consumer goods production, they had to retool to meet war production needs. Each of them found markets for its products, and the war encouraged them to expand their product lines in ways that would serve their needs in the postwar economy.

Owens-Corning Fiberglas, the youngest of the companies, saw the war as an opportunity to expand research and production into new products with a guaranteed market. Insulation produced by the company was used by the military in naval ships and airplanes because it was lightweight. Owens-Corning estimated that using its lightweight insulation in a battleship saved 60 tons, allowing the ship to carry additional fuel for six days of duty.[18] The insulation was also used by deep-sea divers. The company also began experimenting with producing Fiberglas fabric for airplane wings and fuselages because it had the advantage of not shrinking or rotting, and it was not impacted by dramatic changes in temperature. By 1944, almost 100 percent of the company's production was war-related, and it had expanded beyond its Newark, Ohio, plant to plants in Ashton, Rhode Island, and Huntingdon, Pennsylvania.[19] Owens-Corning Fiberglas was so successful that in 1944 *Forbes* magazine featured a story on Harold Boeschenstein and the company's research-and-development efforts.[20]

Owens-Illinois also expanded its product lines to meet the needs of the military. Its glass containers were used to store food products and hermetically sealed blood plasma. Even its beer bottles were promoted for their support of the war effort. A publication produced by the company in 1943 noted that "England found, during the blitz, that beer was a most important element in civilian morale."[21]

Other O-I products used in the war included cardboard shipping boxes produced by O-I's Complete Packaging Service, Hemmingway Insulators used on telephone and telegraph lines, and even Libbey Safedge glassware. The company provided nearly one million glasses for the war. Specific made-to-order glassware such as radio and x-ray tubes was produced by the company as well.

Owens-Illinois Can Company expanded the company's war production beyond glassware, making cans for petroleum products, chemicals, mortar casings, mess kits, and ammunition cartridge boxes. Other specific products made for defense could only be hinted at because of the sensitive nature of the items. These included gauges, levers, bushings, and gears. Most of these were developed by the O-I General Engineering Department, which also produced molds used for making bomber gun turrets at the Alton, Illinois, factory. The company's war production efforts earned it the Army-Navy "E" Flag for "Excellence" several times during the war at its various plants.

As an indication of how the war impacted the company, its net sales escalated from $89 million in 1940 to $174.5 million in 1944, although its net profits during this time increased less than $1 million because of wartime price controls.[22] Many Owens-Illinois employees paid the ultimate price for service in the war. The company's 1943 annual report listed 25 "Gold Stars," the honorary name given to employees who were killed in the war.[23]

Because of his service in Washington and his trip to England in 1941, John Biggers recognized the need to retool Libbey-Owens-Ford for wartime production, especially since the war meant it had lost one of its largest markets—windshields for General Motors. Automobile glass made up 60 percent of the company's production before the war. Biggers and Vice President D. H. Goodwillie described the company's philosophy in a special message to all L-O-F employees sent in December 1941: "We must do everything possible to convert parts of our plants and to use our organization in the fabrication and manufacture of products needed for the defense of America and for the defeat of our enemies. Our company, our management, our workers, and our plants are enlisted for the duration of the war. . . . How far we may be compelled to curtail glass-making operations, we do not know. How much defense work we can secure and fit into our factories and our organization, we do not know. But we do know that these wartime orders to stop production of automobile glass means immediate lay-offs and shutdowns which we deeply regret, especially at this Christmas Season. Therefore, to the task of securing work, of planning efficiently, and of training our organization in new types of production, *we and our associates dedicate ourselves* as we enter the tremendously important year of 1942."[24]

Perhaps more than any other Toledo glass company, L-O-F committed significant research efforts to wartime products. A new Optics Control and Development Laboratory was constructed to make products like pilot goggles and tank prisms, which were desperately needed by the government since access was cut off to the established German precision optics industry. The automobile glass production line was retooled to produce aviation glass for canopies and windshields. This required the company to figure out how to bend glass without distorting it, a problem that L-O-F engineers solved in just four days. The company invested in the development of bulletproof glass, expanded its Thermopane glass production for applications where insulated glass was required, and began producing Plexiglas. The company's Plaskon Division, which it had purchased in 1940, even supplied the raw materials for the buttons for military uniforms.

In order to produce all of the products needed for the war effort when many men were not available for the labor pool, L-O-F, like many manufacturers, employed women in its factories. The myth of Rosie the Riveter has been romanticized in recent years, but the reality was that women working for factories like L-O-F were generally paid less, treated poorly, and assumed to be temporary workers who would leave at the end of the war.[25] The women employed by L-O-F were automatically members of the CIO-Federation of Glass, Ceramic, and Silica Sand Workers of America, but they were not considered equal to their male counterparts. To determine wages, work at the Rossford plant was divided between "heavy work" and "light work." "Light work" paid less, and was given to women.

This discrepancy in treatment and wages came to a head at L-O-F's Plexiglas plant, which had begun to hire large numbers of women in 1942. The women contended they were doing the same jobs as men, and should be paid the same. Women were elected to union offices at the plant, and pushed for equal pay. The company continued to argue that the women were not replacing male workers (which would have supported their contention of equal pay) but were only temporary employees. Similar issues developed at the Rossford Edging Department, which was paying women a third less than men. The case was taken to the War Labor Board, and while the union and the company awaited a decision from this board, the women organized an unauthorized strike.

Strikes were a particularly sensitive issue during the war because they could impact the ability of the country to sustain its war efforts. The War Labor Board eventually sided with the women, but stated that they should receive slightly less than men because of the differences in their jobs and productivity. But L-O-F refused to comply because it said the women were producing inferior products and at a slower pace then their male counterparts. Some 270 women walked off the job in August 1943, angry that young, newly hired boys were making more than they were.

The company retaliated against the unauthorized strike by dismissing 120 of the women. Most were eventually allowed to come back, but the struggle clearly impacted the morale of the women. The company continued to believe that the strike was not only unauthorized, but unpatriotic. The union saw little problem with the discriminatory pay because it continued to see the women as only temporary workers who only got their jobs because of the national emergency, and the union assumed they would return to their domestic roles once the war ended.

There were many other questions for the company about what might be in store in the postwar years. Libbey-Owens-Ford estimated it invested $53 million in equipment for the war, and it was uncertain whether any of this investment would pay off for peacetime production.[26] As for 11 of the Rossford plant men who fought in the war, there would be no such question about their postwar years—they were killed in combat.

Postwar Toledo

The events that occurred between 1939 and 1945 made most forget the optimistic theme of the New York World's Fair—"Building the World of Tomorrow." During the war years, tomorrow was pushed to the background as the world focused on survival from day to day. Rebuilding after the devastation of Europe, the nuclear destruction of Japan, and the deaths of millions seemed overwhelming.

In cities like Toledo, there was concern about how to move the economy from one driven by war production to peacetime production while avoiding another cataclysmic depression. Some movers and shakers considered the end of the war as an opportunity to remake Toledo and create an even greater future—perhaps one that would allow it to finally become the "Future Great City of the World," as Jesup W. Scott had dreamed back in 1868. After all, Toledo had not just survived the war and the worst depression of modern times, but its major economic driver—the glass industry—actually flourished. Research and development underwritten by the war effort had resulted in entirely new product lines. All four of the glass companies were poised to retool for a consumer-driven market that might bring new levels of prosperity to the city.

Even before the war ended, planning for postwar Toledo began. In 1944, the Toledo-Lucas County Plan Commission issued what was called "A Preview of Planning for Toledo and Lucas County" titled *What About Our Future?* Its introduction stated the angst of city leaders. "Now the war has curtailed construction and raised the specter of unemployment after it is won. Plans are urgently needed."[27] The study identified major areas that required careful planning to bring prosperity to the city, including neighborhoods, the downtown, public buildings, primary streets, industrial development, recreation, the Port of Toledo, an airport, and the overall need for capital improvements in the city. The commission proposed to produce a master plan in 1945 that would address each of these areas.

Paul Block Jr., who became publisher of the *Toledo Blade* in 1942 after the death of his father, also worried about the prospects for postwar Toledo. In an effort to attract new investment to the city, in 1944 the *Blade* purchased advertisements in several national trade journals to promote the city.[28] With the *Toledo Blade* prominently featured in each, the ads called attention to the art museum, university, port, public library, as well as the city's industries, including its glass manufacturers. The ads appeared in *Advertising Age*, *Editor and Publisher*, and *Women's Wear Daily*, among others.

To help guide the city's postwar development, Block looked back to the 1939 New York World's Fair and its most popular exhibit, General Motors' Futurama. Futurama, the utopian vision of the future designed by Norman Bel Geddes, would inspire a similar exhibit in Toledo in 1945 called "Toledo Tomorrow," underwritten by Block and designed by Bel Geddes himself. Bel Geddes agreed to the project because of his personal connection to Toledo. He had been born in nearby Adrian, Michigan, and his first wife, Helen Belle Sneider, was a Toledo socialite.

Bel Geddes's career had taken several interesting turns. Educated at the Cleveland School of Art and the Art Institute of Chicago, he began as an artist and theatrical set designer, designing sets for Broadway plays.[29] In the 1920s he moved into industrial design. The firm he founded sought to make products more efficient and aesthetically pleasing. He was an architectural consultant to the 1933 Chicago World's Fair, and that year his book *Horizons* was published, showing modern designs for most household and consumer items. Toledo Scale was prominently featured in this book, including depictions of streamlined scales and a proposal for a new headquarters building to be built on Telegraph Road in north Toledo shown as a three-dimensional model. In 1936, Shell Oil hired Bel Geddes to design advertisements that would showcase new ideas to improve traffic flows, and he made a model city as part of his design. This effort led him to General Motors, and to Futurama.

The popularity of Futurama among the millions who saw it at the 1939 World's Fair quickly promoted Bel Geddes as an important urban planner. Part of the attraction of Futurama was that it was both educational and entertaining. Visitors to the General Motors exhibit building were directed to large, comfortable chairs moving on an elevated track. They traveled over the exhibit and viewed it as if riding in an airplane. The 36,000 square feet exhibit showed a detailed scale model of how our country might look in 1960, and consisted of streamlined skyscrapers, farmland, parks, suburbs, airports, and bridges linked by a system of roads called superhighways that could easily accommodate the predicted 38 million automobiles. Each exhibit car had an audio track that explained what viewers were seeing as they traveled down the 1,600-foot ride. At the end of the ride, visitors were deposited into a full-scale model of the exact intersection that they had just seen from above.

Block hired Bel Geddes to create something similar for Toledo. Block's motives for the "Toledo Tomorrow" exhibit are unknown, but he was quoted as saying it was a "stunt" meant to spark Toledoans' imagination. Bel Geddes and his company spent 18 months studying the city to identify the unique issues that would need to be addressed for a successful and prosperous future. The guiding principle of the project was that Toledo had to double its population by attracting new industry, "and the only way industry will be attracted is by going to great lengths to offer facilities that no other city is prepared to offer."[30] Block's contract with Bel Geddes, set at $50,000 plus expenses, called for creating a model of the city in the future that "should be inspirational to the people of Toledo within the realm of achievements, provided the city, county, state, and federal governments will furnish the funds for its execution."[31]

When completed by Bel Geddes, the "Toledo Tomorrow" exhibit consisted of a three-dimensional model 61 feet in diameter that depicted a brand-new Toledo at some unspecified time in the future. Its massive size was required, according to Bel Geddes, in order to be as spectacular as possible and attract the public in large numbers. The exhibit consisted of 47 panels, each measuring 4 feet wide and up to 16 feet long, which had been constructed in Bel Geddes's workshop in New York's Rockefeller Center (itself a beacon of urban planning). The panels were shipped to Toledo and reconstructed like a puzzle. Most of the panels were simply one-dimensional painted surfaces depicting land surrounding the downtown. But the central portion of the city was portrayed in a three-dimensional diorama of 5,000 buildings and roads built on a scale of one inch equals 100 feet. The model purported to show the entire horizontal scale of the Greater Toledo area.

Visitors viewed the display by climbing a ramp, giving the exhibit a similar "view from the air" appearance as Futurama without the expense of a track with moving cars. Also like Futurama, an audio track described each panel as viewers made their way around the exhibit. The lighting could be dramatically lowered to show the city aglow at night. When finished, the actual cost of the completed model as quoted in several news stories was $250,000, five times the original projected cost.

While the plan laid out a grand scheme for this new city, Bel Geddes was short on specifics of how it would be achieved. In its coverage of the model on exhibit, the *New York Times* noted that the rebuilding could be achieved gradually, but no projected cost was provided.[32]

According to a promotional booklet available to visitors, the *Toledo Blade* "presents Toledo Tomorrow . . . not as a blueprint for the city's planners and builders, but as an inspiration for future living . . . and Dedicates Toledo Tomorrow to the memory of those men of Toledo who died in this war, and to the future happiness of those men and women who will return."[33] In a front-page article, the *Blade* described the exhibit as "a dramatic innovation in city planning . . . depicting this community as it could be in the future if its citizens elect to follow the recommendation of some of the nation's best planning and engineering minds."[34]

Bel Geddes consulted with several planning experts to develop "Toledo Tomorrow." These included Major Alexander de Seversky, an expert on aviation; W. Earle Andrews, a highway engineer; Col. Henry M. Waite, a railroad transportation expert; and Geoffry N. Lawford and O. Kline Fulmer, architects of housing projects. Together, the experts came up with six major innovations for "Toledo Tomorrow." These included the nation's first terminal to unite air, rail, and bus transportation; a network of airfields, including a downtown passenger airport that would not be deterred by surrounding high-rise buildings; the consolidation of railroad lines; a system of "congestion-proof" express highways; a beautified riverfront with heavy industry moved to the area of Maumee Bay; and planned housing communities that were efficient and friendly places to live.

While the plan purported to present a completely new city, in reality it was largely a plan for improving transportation. It took advantage of Toledo's location by suggesting highways that would connect the city with Cleveland, Detroit, and Chicago, creating airports for commercial and private planes, and developing the port for shipping. It included little on how it would improve the quality of life for Toledo's residents, however. Only one section of the accompanying booklet on the topic of housing directly addressed the daily needs of residents. The plan called for slum removal in order to reduce crime and improve public health, noting in this polio-panicked era the need to reduce germs that "recognize no neighborhood barriers." New housing was to be built around cul-de-sacs that would prohibit cars from traveling through neighborhoods, and that would create parklike living areas, even downtown. Only two brief paragraphs noted the need for more recreational areas.

Like its predecessor Futurama, little of "Toledo Tomorrow" ever became reality in the postwar years. A master plan developed by the Metropolitan Planning Committee of the Toledo Chamber of Commerce issued in 1945 mentioned the exhibit, and congratulated Block for sparking Toledoans' interest in city improvement.[35] Some of the plans proposed by the Chamber drew heavily from "Toledo Tomorrow," especially a proposal for inter-regional highways that followed the same routes suggested by Bel Geddes.

There is some evidence that the exhibit—and the *Blade's* continual coverage of the exhibit—did stimulate pride by residents in their city. In 1946, the exhibit was used to help sell a municipal income tax. In an editorial, the *Blade* called on residents to make some of the elements of "Toledo Tomorrow" a reality. "Granted that Toledo Tomorrow was a visionary project so vast in scope that its attainment stretches into the distant future, there are many things to be done at once which can start us on our way. There's a new union station to be built. There's the present airport to be improved and a new one to be established."[36] The editorial also noted plans to construct two highways through the city, one going north-south and one east-west. In April of that year, voters in Toledo approved the income tax, one of the first municipalities in the country to enact such a tax to support improved city services. The *Blade* took some credit for this, stating that "the spirit of Toledo Tomorrow translated into action through the amazing loyalty of the citizens of Toledo Today, inspired by the vision of a future city of still wider opportunities for all people."[37]

Toledo's Glass Industry Retools for Peacetime

While the war pulled Toledo's glass companies out of the Depression, the question for their leaders was how to take the research that produced gun turrets, submarine insulation, and periscope prisms and produce goods in demand by 138 million consumers who had been frustrated by a decade of economic depression and war. As John Biggers noted in a radio address in August 1945, "The guns are silenced. Our enemies are defeated. The nation's wartime production job is done. But a new challenge looms over the horizon of the dawning world peace. It is a challenge to government, to labor, to industry, and to each individual to develop a sound peacetime economy. It is a challenge that must be met. And to do so successfully will require the same ingenuity, the same courage, and the same kind of cooperative effort that enabled us to conquer the production battles of the war just ended."[38]

For one of the leaders of Owens-Illinois, the wartime management had a personal cost. Preston Levis was called to fill in as president of the company for his cousin William while William was in Washington working for the War Production Board. Preston not only had to run O-I, but also was involved in day-to-day operations of Owens-Corning Fiberglas because Harold Boeschenstein was also serving in Washington. All had left him physically and emotionally exhausted. Preston experienced a mental breakdown at the end of the war that led to a yearlong stay at a sanitarium.[39]

Owens-Illinois focused its postwar bottle production on Durglas, and constructed a large Duraglas Service Center in downtown Toledo in 1947.

The service center was a renovated building that featured Insulux glass blocks extensively. It housed a one-stop customer service and sales center where 9,000 Duraglas bottles and molds were displayed. Customers could talk with graphic designers about bottle decoration and labeling, and could easily schedule production. The one-way beer bottle was also a hot product for the company in postwar America.

In 1946, Owens-Illinois expanded its product line again through acquisition, this time by purchasing Kimble Glass of Vineland, New Jersey, for $18.5 million. Kimble Glass was a producer of scientific and technological glass. Its major growth occurred during World War I when the supply of German laboratory glassware was halted and the United States found itself without domestic producers of medical and scientific glassware. In 1922, the company developed the first automatic machine to produce test tubes, and prices dropped so dramatically that test tubes were simply disposed of after one use rather than sterilized and reused. The company's products were again in great demand during World War II, especially products like medical syringes and chemical containers. To keep up with demand, Kimble Glass expanded production to Tiffin Glass in Tiffin, Ohio, and to Libbey Glass in Toledo. With a postwar boom in scientific research that was largely funded by the federal government, Owens-Illinois felt the company had great potential. As the O-I annual report of 1946 noted, "The products manufactured by the Kimble Glass Division constitute a profitable line of glassware in increasing demand by many rapidly growing industries occupying important positions in the Nation's economy. With the application of our Company's services, large research facilities, and mass production techniques, we hope to further enlarge this field, and to better serve the needs of our customers."[40]

After the war, Kimble Glass used its production facilities that had made test tubes to produce tubes needed for radios. When television exploded onto the scene, Owens-Illinois—which was the first company to mass-produce television picture tubes—transferred production of the product to a Kimble factory in Columbus, Ohio.

One of the products that Owens-Illinois developed and sold in the postwar period would later cause it considerable difficulty. It was called Kaylo insulation, and it was initially sold for military applications. O-I began large-scale production of the product for the consumer market in 1947. Kaylo was fire-resistant insulation that was made using asbestos fibers. In 1948, O-I felt so strongly about the potential for Kaylo that it created a wholly owned subsidiary for it and Insulux glass blocks, and called the company American Structural Products.[41]

Owens-Illinois also owned a company that was outside of its usual product line—the Owens Brush Company. The company manufactured toothbrushes, hairbrushes, and toilet brushes. The company had begun in Toledo as the Toledo Automatic Brush Machine Company in 1921, founded by inventor Conrad Jobst. Jobst invented a machine to attach bristles to a plastic injection-molded handle, creating the first practical, sanitary toothbrush. The Owens Bottle Machine Company had first affiliated with the company by creating glass tubes used as packaging for the toothbrushes in order to show consumers that they were sanitary. Owens-Illinois had acquired all of the assets of the company in 1944, and changed its name to the Owens Brush Company in 1945.

Libbey Glass quickly expanded its retail glassware line to meet the consumer needs created by the postwar housing boom. Families moving into the fast-growing suburbs required tumblers and tableware. Libbey's lines followed popular home decor trends. The company employed Freda Diamond, the New York design and retail store marketing consultant, and she developed many of Libbey's most successful products during her 35 years with the company. One popular postwar product was the "hostess set," which included matching cups and plates that were a must for proper entertaining. They came prepackaged, allowing for easy display and marketing in department stores. The company also promoted its "heat treated" glassware, which was more durable. One company that used Libbey exclusively was Coca-Cola, with Libbey producing its trademarked shaped glasses that became recognized worldwide. Three factors guided Libbey's postwar product line: an emphasis on patterned glass with promotional themes, glassware sold in gift sets, and a national advertising campaign for its products.[42]

The investments made by Owens-Illinois to meet the demands of the postwar economy—both in its Owens-Illinois product line and in its Libbey Glass line—were overwhelmingly successful. Profits increased from $8.8 million in 1946 to $24.3 million in 1950.[43]

Libbey-Owens-Ford also saw a major expansion in the postwar years. Its Thermopane glass became a major influence in the architecture of the period. New suburban housing developments, built where land was cheap, allowed for one-story "ranch" houses. The long, low architecture incorporated large picture windows made of Thermopane insulating glass. To promote the energy efficiency of large windows, L-O-F developed a "solar home" marketing campaign.[44] The company purchased a device called a Solarometer from scientists at the Massachusetts Institute of Technology, and began showing it off at trade fairs in 1949. Architects were encouraged to use the device to track the sun's path over any location so that they could orient their new homes correctly. As John Biggers described the solar home concept, "Such houses are properly oriented to take advantage of the warmth, cheerfulness and health-giving properties of the sun. They are built to allow the sun's rays to enter the house through wide glass areas during the winter months, thus actually contributing to the heating of the home. In summer, when the sun is high in the heavens, overhanging eaves shade the house, keeping out the sun's rays during the warm months."[45] In another marketing ploy, L-O-F hired several nationally known architects to design solar homes in various places throughout the country. The solar home built in Illinois was designed by John Lloyd Wright, son of famed architect Frank Lloyd Wright. Thermopane was so successful that in 1946 a new plant was opened in Rossford that produced only Thermopane glass. Despite an eight-fold increase in production capacity, orders for Thermopane outpaced production.

The company's expansion into mirrors with the purchase of Liberty Mirror Works of Brackenridge, Pennsylvania, in 1943 also helped postwar profits as the housing market boomed. The company produced numerous promotional campaigns and published advertisements in the popular magazines of the day that showed new ways to incorporate mirrors into home interiors.

Not only were L-O-F products needed for housing, but the new United Nations building constructed in New York City in 1949 featured 5,600 windows made in Rossford. The company also went back to its profitable automobile glass production, helping to meet pent-up demand by consumers for new cars that had been delayed by the war.

One postwar issue for the company concerned its part ownership of Nippon Glass in Japan. After the attack on Pearl Harbor, all communication was cut off with the Nippon leadership, but L-O-F officials were assured that the company's shares in Nippon had been deposited in a bank and were safe.[46] Biggers expressed surprise to find when the war ended that dividends on L-O-F's stock in the company that accrued during the war were paid to the bank account. "Those of use who knew and admired individual Japanese were saddened by the war between the two countries and never harbored any feelings of bitterness but were highly pleased by the treatment accorded American interests during the years when the relations between the two countries were broken," Biggers later recalled.[47]

But when the war ended, Biggers was alarmed to learn that General Douglas MacArthur and his hand-picked economic adviser to Japan thought it necessary to weaken Japanese industry by dismantling successful large prewar companies, including Nippon Glass. Using the connections he had established during his service with the Office of Production Management, Biggers attempted to get MacArthur's orders overturned. While he fell short of stopping the implementation of the orders, Biggers was successful in getting MacArthur to withdraw an order to liquidate Nippon. Because of Biggers's work, he and the other L-O-F leaders maintained friendly ties with Nippon leaders, and worked with the company to expand production after the war. In 1950, Biggers helped the company build a new factory, and also introduced Thermopane to its production line. Biggers traveled to Japan after the war to meet personally with the Nippon leaders.

Like Owens-Illinois, Libbey-Owens-Ford expanded rapidly in the postwar period. Profits went from $9.9 million in 1940 to $24.7 million in 1950.[48]

Owens-Corning Fiberglas was probably in the most tenuous position of the Toledo glass companies following the war. Nearly 100 percent of its production had been war production. Because it was the newest company, it was also the least established with the consumer market. When war broke out, the company had to shelve over 300 research projects in the works to produce consumer products.[49]

Most of the products it did market after the war were aimed at industrial applications. These included insulation for appliances, pipes, aircraft, busses, and refrigerated railroad cars. Its major consumer product push was in the area of Fiberglas fabrics. These fabrics were sold for both their insulating properties (particularly useful in postwar housing construction that featured large picture windows) and for their fireproof quality. Fiberglas curtains were sold to many public auditoriums. In addition to being fireproof, the curtains were marketed as being resistant to mildew, shrinking, and rot. The curtains could be washed in a standard washing machine and rehung without ironing. One of the biggest obstacles to the home consumer market was the fact that the company struggled to find ways to produce anything other than white cloth because the Fiberglas fabric would not accept dyes. Eventually, the company did develop a method for producing patterned Fiberglas cloth. For his efforts leading Owens-Corning Fiberglas, Harold Boeschenstein was honored with a cover story in *Business Week* magazine in December 1948.[50]

Legal and Labor Troubles

Three legal cases would rock the glass industry in the years immediately following the end of the war.[51] One dealt with glass container manufacturers, one with flat glass manufacturers, and one with glass fiber manufacturers. The cases were brought by the U.S. Department of Justice, which accused the industries of antitrust actions resulting in monopolies and price fixing.

The first case, *United States v. Harford-Empire, et al.*, dated back to 1938 when the Temporary National Economic Committee, a New Deal committee that sought to break up monopolies, began investigating price fixing in the glass container industry. At the time, Hartford-Empire, a company largely owned by the Houghton family of Corning Glass, made bottles using the gob-fed production method. Owens-Illinois made bottles using the suction method. But both companies had been cross-licensing their production methods for years, and by 1938 the two controlled 97 percent of glass container production.

The investigation by the Temporary National Economic Committee led to the Justice Department filing suit in 1939 in the District Court in Toledo against Hartford-Empire, Owens-Illinois, and the Glass Container Association (a trade organization that, the government contended, encouraged the monopolistic practices). In 1942, the court handed down a sweeping judgment that restrained the companies from unfair licensing practices and price fixing, and called for the liquidation of Hartford-Empire. The companies appealed the case to the U.S. Supreme Court, which ruled in 1945 that the companies had indeed participated in monopolistic practices. But the court also set aside much of the draconian settlement provisions that had been decided by the district court. In the end, no fines or penalties were assessed, only court costs. But for Owens-Illinois, that was $2.87 million, not to mention the legal fees that the company had accumulated in seven years of fighting the case. The Supreme Court did agree that the Glass Container Association should be dissolved, but the same year the Justice Department allowed a new organization called the Glass Container Manufacturers Institute to be formed.

In 1945, a nearly identical case was brought against flat glass producers. The case, the *United States v. Libbey-Owens-Ford Glass Co., et al.*, said that flat glass producers had also been using patents and licensing to control the industry. The outcome of the case, which was settled without going to trial, made the patents of the companies, including L-O-F, available to anyone willing to pay a reasonable royalty to the company for use of the patents.

In 1947, *United States v. Owens-Corning Fiberglas Corporation, et al.*, was filed by the Justice Department. This case dealt with the unique structuring of Owens-Corning, and the continued involvement of Owens-Illinois and Corning Glass in the operations of the company. The Justice Department claimed this was an attempt to monopolize the glass fiber industry. The case was decided in 1949, and required that O-I and Corning end their close relationships with Owens-Corning, although both continued to hold stock in the company for many years to come. Owens-Corning was also ordered to make its patents available for a reasonable royalty fee, and provide technical assistance to those wanting to license the patents.

In addition to concern over business monopolies, another legacy of the New Deal was an invigorated labor movement, including in the glass industry. During the war, labor strikes were contained by the War Labor Board and a sense of patriotism. Strikes meant that goods needed by the men on the front were not being produced, and corporate leadership warned of dire consequences for these men if strikes were called.[52] This attitude was displayed during the strike by the women at Libbey-Owens-Ford in 1943, when the company called those who walked off the job because of a discriminatory pay structure unpatriotic. Similar charges were leveled against L-O-F workers in 1943 and 1944 who were angered by company meddling in shop-floor rules. Union workers walked off the job, but not without expressing concerns about how the strike could be perceived as unpatriotic. The regional War Labor Board heard the workers' case, and ordered them back to work.

In 1944, the heads of L-O-F and Pittsburgh Plate Glass and leaders of the CIO-Federation of Glass, Ceramic, and Silica Sand Workers of America met to begin negotiations on a new contract. The companies demanded new controls over issues like promotions, which threatened the union's seniority system. The union demanded an increase of 17 cents per hour. Because the war was still going on, the issue went to the War Labor Board, which rejected the union's wage increase request but suggested the inclusion of a contract reopener over wages when the war ended and controls over both wages and production came to an end.

What the 1944 contract negotiations taught the union was that the period of peace and openness between labor and management at L-O-F was ending, and a new period characterized by the company inserting itself into union issues was beginning. The "cooperation clause" that had been negotiated when L-O-F first recognized the Federation as the bargaining agent for its employees—which stated that the company would "continue the practice of cooperating with the union to the best interest of both parties"—was no longer in effect.[53]

The company heads and labor leaders met again in May 1945 in an attempt to negotiate a new contract that would set the stage for postwar production. But while the war was winding down, and victory in Europe would come soon, the union was again labeled as unpatriotic when it served a 30-day strike notice because of a lack of progress in negotiations. While some Toledo labor leaders, particularly the powerful Richard Gosser of the United Automobile Workers, opposed a strike, the L-O-F workers of Local

9 of the Federation voted to strike on June 14. While the union had already yielded on many of its demands, it still asked for a wage increase of 10 cents per hour. The company countered with an offer of 8 cents.

In June, the War Labor Board ordered the workers back to work, and stated that the board would take up the matter later. When the union and the companies met again in October, the landscape had changed dramatically. The war had finally ended, and the federal government had begun to remove some of its controls over production and wages. Most importantly, striking could no longer be seen as unpatriotic. The Federation now demanded an increase of 25 cents per hour. In October, workers at the L-O-F Ottawa, Illinois, plant closed down the plant in an unauthorized strike. John Biggers was quoted in the Toledo newspapers as saying the strike was an "outlaw strike," and the company would refuse to negotiate until it ended.[54] The Federation urged both L-O-F and Pittsburgh Plate Glass to reconsider their position, but when they would not, the Federation shut down production at all flat glass facilities in the country. The strike idled 15,000 workers, 4,000 of them at L-O-F's three Toledo plants.[55]

The strike continued through December. The union was frustrated that the companies would not negotiate, and the Department of Labor finally appointed a conciliator in an attempt to bring the two sides together. Negotiations resumed in late December, and in January 1946, the strike ended. The unions agreed to an immediate 10.7 cent per hour increase, and other issues were sent to arbitration. The result of arbitration was that the companies reaffirmed the rights of the Federation to negotiate for the workers in the flat glass industry, while the union agreed to stop unauthorized strikes. The agreement brought relative peace between labor and management, just in time for postwar production.

The Toledo Industrial Peace Board had been organized after the major strikes of the Depression era (including the violent Toledo Electric Auto-Lite strike) in an effort to bring peace between labor and management and improve the business climate in Toledo. In the years the board had existed, it had helped to settle 72 strikes and 74 threatened strikes in Toledo. It had also brought national attention to the city as a place where labor and management could work out their differences peacefully.[56]

The board was suspended during the war because labor relations were assumed by the federal War Labor Board. It was reconstituted at the request of Vice Mayor Michael V. DiSalle and the city council after the war as the Toledo Labor-Management-Citizens Committee. The management side of the committee was represented by John Biggers. William Akos of Local 9 of the Federation of Glass, Ceramic, and Silica Sand Workers was appointed an associate member of the committee in 1947 with the endorsement of the Toledo Industrial Union Council. Through this appointment, the flat glass workers confirmed their interest in postwar labor harmony.

Toledo's Yesterday and Tomorrow

In the 1940s and 1950s, Toledo and its glass companies celebrated their long history together with several commemorative events. In 1946, Owens-Illinois and its Libbey subsidiary donated the massive cut-glass punch bowl made for the 1904 St. Louis World's Fair to the Toledo Museum of Art. The punch bowl had been in a Libbey storeroom since the fair had ended 42 years before. John Denman, the cutter who had made the punch bowl, was honored at the ceremony marking the donation. In turn, the museum honored the glass industry in 1951 with an exhibit marking the museum's 50th anniversary entitled "Art in Crystal—A Historical Exhibition of Libbey Glass 1818–1951." The exhibit included the St. Louis punch bowl and some of 250 other priceless cut and engraved pieces donated by Owens-Illinois to the museum that year. But in the same year it celebrated its history, Libbey also demolished the original two factory buildings at the Ash Street site that had been built by Edward Drummond Libbey in 1888.

Owens-Illinois celebrated another aspect of its heritage in 1959 with events in 34 cities marking the centennial of the birth of Michael Owens. Commemorative glass plaques were hung in all O-I factories. An article in the *Blade* about the celebration noted that in 1959 sales at the three companies that bore Owens's name topped $1 billion and employed more than 55,000 people.[57]

The glass companies dominated Toledo's booming postwar economy. The city began to refer to itself as the "Glass Center of the World." But as for the postwar vision of "Toledo Tomorrow" laid out by Norman Bel Geddes and promoted by *Blade* publisher Paul Block Jr., little came to fruition. A new Union Station was built in 1950 on the site of the old railroad terminal that dated back to the nineteenth century, but it was not connected to air or bus transportation, as "Toledo Tomorrow" predicted. No downtown airport (or the three other airports proposed) was ever built. In 1952, six Toledo corporations, including Owens-Illinois, purchased land for an airport, but it was located in Swanton Township, some 15 miles east of Toledo. The companies then sold the land back to Lucas County at cost, and the new airport was dedicated in 1954. Some postwar housing developments were built around a cul-de-sac plan, such as Lincolnshire in west Toledo. But this model of neighborhoods was largely relegated to the suburbs, where land was plentiful and cheaper, allowing for less densely populated neighborhood planning.

Perhaps the most important impact of "Toledo Tomorrow" was on its patron, Paul Block Jr. From that point on, Block took an active interest in Toledo city planning, particularly the development (and redevelopment) of the downtown. And because Toledo's economy was driven by the glass manufacturers, these companies would play a major role in future plans for developing the downtown.

Sometime in the 10 years after the "Toledo Tomorrow" exhibit closed, the massive model was discarded because it became too costly to store. But the story of how Paul Block Jr.—in partnership with the corporate leaders of Toledo's glass industry—sought to create a real-life "Toledo Tomorrow" is, in many ways, the story of the city itself in the second half of the twentieth century.

SEVEN

Expansion, Energy, and the Environment

> We all have a stake in Toledo's destiny. And if each of us will accept some personal responsibility for helping to meet Toledo's challenges, the great accomplishments of the Sixties will prove to be merely the first step along the road to true progress in our city.
>
> —Edwin Dodd, chief operating officer of Owens-Illinois, in a speech to the Toledo Chamber of Commerce, January 1970

The three decades following the end of World War II were periods of sweeping growth for the glass companies of Toledo that would irreversibly change their futures—and the future of the city they called home. Through the acquisition of other companies and the development of new product lines, all were able to diversify beyond glass, helping to insulate them from the sort of disruptive events that had shaped their early years. They also expanded globally, establishing themselves as international companies with subsidiaries around the world. Two of the glass companies would undertake major building projects for new headquarters in Toledo, and a third would begin planning for an even greater vision for the city's downtown.

But not all of the developments were positive. The rising cost of energy had a devastating impact on an industry that uses large amounts to manufacture most of its products. The environmental movement also impacted the glass industry, particularly Owens-Illinois, inventor of the one-way beverage container that came to be seen as wasteful and destructive to the country's landscape. Lastly, high interest rates on money borrowed for expansion and acquisitions, and periodic economic recessions throughout the decades, sowed the seeds for struggles that would impact all of the companies in the 1980s and 1990s.

A New Leader and a New Headquarters

While it was one of the largest flat glass companies in the country in the 1950s—second only to Pittsburgh Plate Glass—Libbey-Owens-Ford still rented space for its headquarters on the twelfth floor of the Nicholas Building in downtown Toledo, as it had done from the time of its founding in 1929. The offices had been updated in 1941, and renamed the "Glass Headquarters." The renovations incorporated many of the L-O-F's products like Vitrolite and Thermopane into the design so that visitors could see firsthand the products the company sold. But by the late 1950s, the rented headquarters seemed no longer appropriate for a company of the size and stature of L-O-F.

In May 1958, the company announced plans for a new office building. It would be the first major office building constructed in downtown Toledo in 30 years. The site selected was the corner of Madison and Michigan, and the architects Skidmore, Owings, & Merrill were chosen for the modern 14-story design.

The building incorporated a curtain wall construction process that had been pioneered in the early 1950s by the architect Mies van der Rohe, among others. Such construction was made possible by the development of L-O-F's Thermopane insulating windows. The building was constructed of steel girders skimmed with glass panels framed by aluminum. The 6-feet by 10-feet windows of the L-O-F building were unusually large for the time. The outside was coated with what was called Parallel-O-Grey, and the floors were separated by Vitrolux spandrels. The Parallel-O-Grey glass was produced on the company's new twin grinding machine that could grind smooth both sides of a large piece of plate glass at once. Two sheets of the glass were then sealed together to form the Thermopane window, and a coating applied that gave the windows a gray reflective color. The building was described as a "free-standing design, its superstructure overhanging the lobby floor, and its pillar supports resting in the center of the landscaped plaza."[1] Inside, glass was featured in office partitions, and blue glass mosaic tile decorated the first-floor lobby.

The new headquarters building for Libbey-Owens-Ford, built in 1960. (Ward M. Canaday Center for Special Collections.)

The building cost $10 million to build. Its lavishness seemed in keeping with a company that had seen its yearly earnings go from $2 million to over $50 million in the 30 years since it had been founded.[2] The company moved into the new headquarters in March 1960. The construction of the headquarters building coincided with another major expansion—a building adjacent to the company's technical center on Toledo's east side—which brought together all of the engineering aspects of the company.

The new headquarters building also coincided with the retirement of John Biggers. Biggers, who had started working in the glass industry for Michael Owens, led Libbey-Owens-Ford for 30 years. He had seen the company through from its founding to the depths of the Depression and the expansion into military production for the war. It had been Biggers who had secured the company's fortunes by negotiating the exclusive contract to produce windshields for General Motors. He had also helped negotiate peace with the city's labor unions through his leadership of the Labor-Management-Citizens Committee. His work for the federal government in the Office of Production Management and his assignment as special minister to Great Britain before World War II had earned him the President's Medal for Merit. *Forbes* magazine had honored him as one of its "Men of Achievement." He was succeeded in the leadership of Libbey-Owens-Ford by George P. MacNichol Jr., yet another person with a long history in Toledo's glass industry. MacNichol was the grandson of Ford Plate Glass founder Edward Ford.

New Products and Possibilities

Libbey-Owens-Ford continued to develop new products, and was involved in several high-profile projects that earned the company national attention. In September 1951, a new display for the country's Declaration of Independence and Constitution was unveiled in Washington, DC, that featured the documents hermetically sealed in helium-filled Tuf-flex glass developed by Libbey-Owens-Ford and produced at the company's Rossford plant. The display was the result of four years of research conducted jointly by L-O-F engineers and scientists at the National Bureau of Standards.[3] The inert gas enclosure was designed to protect the documents, and filters on the glass blocked out harmful light rays.

In 1954, Libbey-Owens-Ford merged its fledgling fiberglass operation that it had hoped would be a competitor to Owens-Corning with Glass Fibers, Inc., a company in Waterville, Ohio, that had been started in 1944 by Randolph H. Barnard, a former engineer at Owens-Illinois. The new company, known as L-O-F Glass Fibers, remained as a subsidiary of L-O-F. The merger was logical, as Glass Fibers had significant technical and scientific expertise, but lacked enough capital to expand.[4] The Libbey-Owens-Ford fiberglass operation had the necessary capital, but little in the way of research and development. It was hoped that the new company would be able to build on L-O-F's close connection to the automobile industry. As *Business Week* magazine prophetically noted in an article about the merger, "There are some industry observers who foresee the day when glass-reinforced plastic will gain huge popularity as an auto body material."[5] While such a development did happen, it would be pioneered by rival Owens-Corning Fiberglas, not L-O-F. The L-O-F subsidiary was eventually sold, and became Johns-Manville.

Another effort to publicize its cutting-edge products came in the sum-

The encapsulation of the Declaration of Independence in the National Archives between sheets of L-O-F Tuf-flex glass, 1951. (Ward M. Canaday Center for Special Collections.)

mer of 1957 with the opening of the "House of the Future" at Disneyland in California. The space-age design featured a white plastic round house cantilevered out over a central pedestal, resembling a mushroom that had sprouted out of the earth. It was built by Monsanto, the chemical company, to call consumer attention to new building products. The company invited housing industry suppliers to work with it to develop the house that featured imaginative and futuristic uses for products and electronics. Libbey-Owens-Ford provided the insulated windows as well as colorful safety glass screens that were used throughout the house to bring in light.[6] The house featured such gimmicks as a pushbutton telephone, an adjustable bathroom sink, and central climate control that could purify the air and add pleasant scents. While largely a stunt, the "House of the Future" helped L-O-F to continue to market itself as a future-oriented and cutting-edge company.

In 1959, the company received some negative publicity when it was accused by the Federal Trade Commission of using "trickery" in commercials it produced to promote its safety glass in General Motors automobiles.[7] The advertisements had been shown during the popular *Perry Mason* television series and football games between September 1957 and June 1958. The complaint—that photography that claimed to show the clarity of safety glass was actually taken through an open window—was part of the government's effort to rid television commercials of deceptive advertising. It was launched in the wake of quiz show cheating scandals that had rocked the nation. At first, L-O-F denied the charges, but during the second day of testimony at an FTC hearing, the company admitted that the commercial did indeed use an open window. However, the company blamed the error on the commercial's production company.

Two years after the FTC investigation, Libbey-Owens-Ford lost its "cash cow"—the exclusive contract to produce windshields for General Motors that it had held since 1931. GM contracted with Pittsburgh Plate Glass to supply some of the glass for the company's 1962 car line.[8] Neither PPG nor GM would disclose the value of the contract to supply the glass, but it was described as "substantial."

Libbey-Owens-Ford was already experiencing a downturn in sales, with profits declining 20 percent between 1960 and 1961. That year the company switched from producing laminated safety glass to making tempered safety glass in an effort to reduce costs. When laminated glass broke, it was held together by a sheet of plastic sandwiched between the two sheets of glass. When tempered glass broke, it shattered into very small but safe pieces. The new process allowed the company to eliminate more than 1,000 workers from the payroll.[9]

With competition in windshield production, Libbey-Owens-Ford turned its attention to architectural glass innovations. In 1962, it launched its "Open World" advertising campaign that focused on incorporating large expanses of glass in a broad range of buildings that allowed in light while maintaining comfort in all kinds of weather. The company asked several architectural firms, including Samborn, Steketee, Otis & Evans of Toledo, to imagine large glass skyscrapers designed for multiple uses, including offices, apartments, and retail shops. The Toledo firm designed a tall geometric form that utilized high-tensile cables to hang floors as in a suspension bridge, which it called the "Torsion Tower."[10] Henry L. Stojowski of New York designed a "Multistory Prism," featuring precast concrete window casings. Advertisements featuring the designs appeared in several architectural magazines.

In order to reduce the costs of plate glass—which, despite the ability to grind both sides simultaneously remained labor intensive and expensive to make—Libbey-Owens-Ford purchased a license to produce "float glass" from the British firm Pilkington Brothers Ltd. in April 1963. Float glass, as developed by Pilkington in 1959, required no polishing or grinding to create smooth glass. The process used molten glass fed onto a bath of molten tin. By controlling the temperature of the tin, the glass could be slowly cooled as a flat ribbon of smooth uniform thickness. Libbey-Owens-Ford built an entire production facility in Lathrop, California, to produce float glass in 1964, and two years later added a second production facility in Rossford. A third facility was added in East Toledo in 1968. All of the facilities reduced labor costs significantly. Float glass was also used in one of the largest construction applications the company had been involved with to date in 1968 when it supplied panels 9 feet by 30 feet in size for the General Motors building in New York City.[11] L-O-F built special fabricating equipment to produce the large panels.

By the late 1960s, Libbey-Owens-Ford was generally perceived in the business community as a conservative and moderately profitable company based on safe and slow-growing investments. But with new leadership under Robert G. Wingerter in 1967, who was the first corporate leader brought in from outside the L-O-F circle, that would change. Within the first year, Wingerter began to diversify holdings by acquiring companies outside of glass and plastic production. The first major acquisition was the Aeroquip Corporation in 1968, a leader in fluid power that manufactured high-pressure hoses, couplings, and seals. As L-O-F's annual report stated that year, "Acquisition of this fast-growing company, with its variety of promising new fields, provides L-O-F with a major force to broaden its sales base and accelerate its growth. The move provides Libbey-Owens-Ford diversification without sacrificing its strong position in the glass markets."[12] The acquisition also brought Don T. McKone into L-O-F corporate leadership.

To reflect its changing product line, in 1968 the company removed the word "glass" from its corporate logo. In 1969, it purchased another nonglass business, Woodsall, Inc., a plastics molding company. In 1970, it bought Super Oil Seals & Gaskets Limited of Great Britain. The annual report of 1970 attempted to reassure stockholders that the company was not going too afar. Company leaders said they had no intention of turning L-O-F into a conglomerate, but rather would focus on acquiring companies that produced products that were compatible with glass.[13] Another reason for the diversification was the high production costs of glass, especially once energy costs began to soar. The company estimated the per-employee cost of glass production in 1969 at $27,000, while costs in the 500 largest U.S.

industrial companies that manufactured other products was nearly half that, at $14,000 per employee. To further cut costs, in 1971 the company closed it long-standing Shreveport, Louisiana, plant because of what it said was labor's unwillingness to reduce production costs at the factory.[14] The closing resulted in the elimination of 460 jobs.

The company continued converting its entire glass production to the float glass process, investing over $100 million in the conversion. Robert Wingerter called this the largest capital investment in the company's history. The changeover came just in time for another high-profile project for Libbey-Owens-Ford—supplying the glass for the construction of the world's tallest buildings, the World Trade Center in New York City, beginning in 1970. The buildings were designed by Minoru Yamasaki & Associates of Troy, Michigan, a firm that would play a large role in Toledo's development in the 1980s. The World Trade Center contained enough glass for 3,600 private homes, and cost $650 million to build.[15] Libbey-Owens-Ford had to develop a new system to bring the glass to the construction site located in the center of densely populated Midtown Manhattan. It developed reusable steel containers to transport and store the glass, eliminating the disposal problem that would have resulted from using wooden crates.

The tempered glass used in the project was several times stronger than the glass usually installed in skyscrapers because the glass walls were self-supporting. The upper floors used Bronze Tuf-flex glass to reduce air conditioning costs, and plate glass was installed in the lower floors. The glass could be installed from within the buildings' structure, eliminating the need for glazers to hang on the outside. As the *Blade* reported, "Even [the buildings'] detractors must agree that it is a monument to engineering and architectural skill. And tourists from all over the world are adding to their lists of must-see attractions the buildings that L-O-F likes to call 'stairways to the stars.'"[16]

In 1970, L-O-F used its technology to improve the experience of concertgoers attending Toledo Symphony Orchestra performances. At the request of avid symphony patron Dr. Edward H. Koster, the company built a glass lid for the orchestra's grand piano. As Koster explained in a letter to the company, "The lid on a piano is an important factor in projecting the sound out into the concert hall, however it provides a big black blob of a sight barrier for the audience. It completely occludes the podium and, depending on the particular seat in the auditorium, most of the orchestra from vision."[17] The solution L-O-F developed was two pieces of quarter-inch tempered safety glass laminated together. The lid could fit both of the Toledo Symphony's grand pianos, although it weighed twice as much as a conventional lid.

When Arthur Fiedler, the conductor of the Boston Pops Orchestra, visited Toledo in 1974 for a performance and noted the glass top, the company produced a similar one for his orchestra, and presented it to him as a gift.[18]

In December 1973, John Biggers died at the age of 85. Robert Wingerter paid tribute to Biggers's leadership of the company. "No person felt neglected or unimportant in John Biggers' presence. His kindness and thoughtfulness touched everyone, from the highest to the very lowest in his host of friendships. . . . As a tremendous American, he reflected the very highest kind of responsible business and civic leadership."[19]

Owens-Corning Fiberglas Matures

As the youngest of Toledo's glass companies, Owens-Corning Fiberglas experienced a growth spurt in the years following World War II. The company finally found markets for its products, particularly the new area of fiberglass-reinforced plastics. Under the leadership of Harold Boeschenstein, the company grew literally to new heights with the construction of a headquarters building in downtown Toledo that would be the tallest building in the city until it was surpassed by Owens-Illinois. It was also during this time that the Fiberglas made by the company became distinguished by its pink coloration, a step taken in 1956 by adding red dye to the binder formula in order to differentiate a finer grade of Fiberglas developed that year from the product made previously. The color pink soon became synonymous with the company.

Boeschenstein's leadership extended beyond his company. During his service on the War Production Board during World War II, he cultivated many high-level contacts in the federal government, including Dwight D. Eisenhower. These relationships pushed Boeschenstein into some highly political roles in the years after the war.

In 1958, President Eisenhower appointed Boeschenstein as chairman of Committee on World Economic Practices. The president asked this special committee to study new ways in which the government and private enterprise could join together to "combat the Sino-Soviet economic offensive and promote free world economic growth."[20] The committee's report, issued in January 1959, noted the increasing Communist economic capacity. In a Cold War warning, the committee stated, "Communists are using their power to influence and control such countries as they can. Communist designs and maneuvers, playing upon the cravings of increasing numbers of people for more and better things, pose a long-term problem of the first magnitude."[21]

The report called for more growth in private enterprise to counteract the state-controlled growth of Communist countries. It also called for increasing foreign aid to developing countries to fight Communist propaganda and influence, but urged that such aid "not be of indefinite duration."[22]

Boeschenstein's service as chair of the committee led to an unusual opportunity to experience Soviet leadership firsthand. In 1959, he was asked to accompany Vice President Richard Nixon to Russia to open the American National Exhibition in Moscow. The exhibit was intended to showcase American ingenuity and entrepreneurship to counter Soviet propaganda about its own industrial might. Also along was another Toledoan—Acting Assistant Secretary of State Foy D. Kohler. Boeschenstein attended the exhibit's opening, where Vice President Nixon was confronted in a "kitchen debate" by Soviet leader Nikita Khrushchev. During what was supposed to be a friendly stroll through a mock-up of a modern American home, Nixon and Khrushchev began to argue over whether the home actually represented better technology than that of the Soviet Union. Khrushchev pointed out that while the house as displayed would cost American workers $14,000, all that was needed to acquire a home in the Soviet Union was to be a citizen of the country. The model home's dishwasher became a particular point of contention between the two, with the Soviet leader claiming that a dishwasher was no competition for the strength of Soviet rockets. Boeschenstein witnessed the event as a member of the audience.

On that trip to the Soviet Union, Boeschenstein also accompanied Nixon on tours of manufacturing plants. When he returned, Boeschenstein was interviewed by the *Blade* about what insights he had gained into the Soviet economy. As might be expected, Boeschenstein expressed his belief that the Soviet plants were not up to the quality of American manufacturing facilities, especially in terms of new technology and safety features. In regards to the Russian people, Boeschenstein reminded the paper's readers, "We must understand that there are 8-million Communist Party members in the Soviet Union who govern the other 200-million Russians. The almost unchallengeable domination under their political system can be appreciated only when you realize that every manager, every administrator of a plant or any other enterprise—a store, a school, a museum, a playground—is appointed by a government, which in turn is controlled by the party 'machine.'"[23] But Boeschenstein also noted that he felt optimistic after meeting so many Soviet citizens. "From Moscow to the heart of Siberia, they gave spontaneous evidence that they like Americans, and the people plainly long for peace and our friendship," Boeschenstein said.[24]

Boeschenstein would continue to advise all U.S. presidents up through the administration of Richard Nixon. On many occasions, President Nixon would send personal notes to Boeschenstein, thanking him for his counsel.

A Tower of Fiberglas

Like Libbey-Owens-Ford, Owens-Corning Fiberglas had rented space for its headquarters in the Nicholas Building in downtown Toledo since 1938, when the company was founded. But Harold Boeschenstein had grander ideas for his company, and in the early 1960s began planning for a new headquarters that would be the tallest building on the Toledo skyline.

Urban renewal was a federal government program begun in 1949 to remove "blight" from the core of cities. With so many people moving to the suburbs in the postwar building boom, the central cores of cities like Toledo were left derelict. Through urban renewal, federal dollars were available to communities to tear down buildings and encourage new development.

In Toledo, one of the biggest urban renewal projects began innocently with a midnight stroll downtown in early 1963 by Mayor John Potter and Toledo architect Orville Bauer. Walking down Summit Street near the Maumee River, the mayor decided that the area either had to be renovated or demolished.[25] Bauer contacted Columbus developer John Galbreath to ask what ideas Galbreath might have for the area, and if he would be interested in investing in the project. Galbreath responded by asking world-renowned architect I. M. Pei to help develop a design for a six-block area bounded by the river, Madison Avenue, St. Clair, and Monroe streets. Mayor Potter expressed his hope that the plans would restore Summit Street and the river to its former stature as a hub of business and commerce.

In July 1963, an opportunity to improve part of the site presented itself when the Charities Foundation, a local philanthropic group that had been founded by William Levis and Harold Boeschenstein in 1937, offered to donate $200,000 to create a park on the site of the old post office to honor Levis, who had died the year before.[26] The city could acquire the building for little money as surplus federal property, and urban renewal funds would pay for demolition. The foundation's money was to be used to landscape the area, with long-term maintenance the responsibility of the city. The officers of the charity noted that donating money to build a downtown park was just the sort of act that William Levis would have done if he were still alive.[27]

Together, the urban renewal project around Summit Street and the Levis park became known as Riverview. Galbreath and Pei actually proposed two phases to the project, with the first phase focusing on the Levis park and a 30-story major office building. If the two had any prior knowledge that the office building would be embraced by Harold Boeschenstein as a new headquarters for Owens-Corning Fiberglas, they did not reveal it. Speaking to members of the Downtown Toledo Associates, a group of downtown business owners, Pei noted that the area for phase 1 was relatively small—only a little over 13 acres. But as Pei expressed, as with any real estate, it was all about location, location, location. "It's so ideally situated that perhaps some of you who are so close to the picture may not see it quite the same way we do. It's situated right adjacent to your 100 percent location for retail, for office buildings, and for hotel operations. At the same time, it is right on the river," Pei noted.[28] He also noted that because of blighted nature of the area, the value of the land was only $800,000 to $900,000. The development proposed would was estimated to add $15 million to the tax base. Mayor Potter formally asked Congressman Thomas Ludlow Ashley for his assistance in gaining urban renewal money for development in January 1965.

But as would be the case with other future downtown development plans, not all of the citizens of the city embraced the project. One anonymous citizen wrote to Mayor Potter that the city was "trying to revitalize the downtown area, so [you] decide to destroy the downtown."[29] In particular, many were disturbed by the plans to tear down the Fort Meigs hotel, which would eliminate one of the few low-cost housing options in the downtown. But it was not only citizens who would delay the project. As architectural plans were fully completed, the cost of the Levis park escalated to $400,000, money that the city did not have.

In May 1965, the city received a $1.3 million urban renewal grant from the federal government to begin the Riverview project. But the project was largely stalled until March 1966, when Boeschenstein announced the intention of Owens-Corning Fiberglas to occupy the proposed office tower.[30] Boeschenstein revealed his plans just as the city was opening bids for the purchase and redevelopment of the project. The project also gained the support of two powerful Toledoans—Stephen Stranahan (president of the Entelco Corporation and a member of the family that founded Champion Spark Plug) and Morris Fruchtman (president of Donovan Steel). Stranahan, working with developer Galbreath, assumed the presidency of the Riverview I Corporation, the entity that purchased the land from the city and oversaw development. By the time construction began, the estimated investment by public and private funds in the project was $125 million.[31]

As quoted in the *Blade*, Mayor Potter expressed his thanks to Owens-

The Fiberglas Tower, ca. 1970. (Ward M. Canaday Center for Special Collections. Used by permission of Owens Corning.)

Corning for kick-starting the project. "This heralds a great day for the future of Toledo, and the people of our city are indeed grateful to see such leadership provided by one of the great industrial citizens. The city is deeply indebted to Owens-Corning for providing this bold, imaginative leadership in this vital area of downtown development."[32] Boeschenstein also won praise from John Galbreath, who called him "one of the leading citizens not only of Toledo, but of the United States."[33] Owens-Corning agreed to pay $1.1 million per year to rent the high-rise office building.

Groundbreaking for what was named the Fiberglas Tower took place on May 1, 1967. At the ceremony, Mayor Potter called the building "a beacon and a guide to point the way to the great Toledo for which we all strive."[34] Stranahan announced at the ceremony that there would be a restaurant on the 28th floor of the building that would allow diners to look out over the city's skyline. Initial plans for the restaurant indicated it would have a Spanish decor to mark Toledo's sister city relationship with its Spanish namesake. The restaurant was to be operated by the Grace E. Smith Company, founded in 1916 in Toledo and operator of the city's landmark Smith's Cafeteria. There would also be a bank located on the first floor as well as retail shops.

In honor of its new headquarters building project, Owens-Corning Fiberglas began publishing a new company newsletter called *Tower News* in

May 1968, replacing its previous publication *Fiberglances.* The publication date coincided with the "topping out" ceremony for the building, marking the installation of the top-floor structure. Boeschenstein used that occasion to express his continued hope that the building would begin a resurgence in the downtown. Stephen Stranahan noted that without such development, the central city business district would become obsolete. "The future is promise," he concluded.[35]

When completed, the Fiberglas Tower was more than just the tallest building in the city. It was a showcase modern office building. Owens-Corning utilized an "office landscape" design, and produced decorative Fiberglas screens to divide office space rather than walls. There was also extensive use of live plants throughout the building—so many that employees had to be educated as to what kind of plants were what and how to care for them. The space was designed by Conrad Zamka, working for the New York design firm of Knoll Associates. He believed that the layout would spark creativity among the employees, and open up the company to spirited and productive discussion. Fiberglas furniture was designed for many of the offices to better show off the company's own products.

The company also began purchasing modern art to decorate the walls of its modern headquarters. A committee chaired by the director of the Toledo Museum of Art oversaw the purchase of paintings that some Owens-Corning employees criticized as too abstract. Each work was carefully placed in the building to set off the color, design, and feel of the wall space, creating a detailed composition.[36] Part of the collection was also displayed in a major exhibition at the Toledo art museum in 1968 prior to the opening of the Fiberglas Tower.

With the completion of the Fiberglas Tower in late 1969, work could finish on Levis Square, the park adjacent to the building. The restaurant at the top of the building was appropriately named "The Top of the Tower," and it opened in the fall of 1970. It quickly became a favorite place for Toledoans to celebrate special occasions.

New Products Fuel Growth

The Fiberglas Tower would not have been possible without the phenomenal growth of Owens-Corning that began in the 1950s. With the end of World War II and its government contracts, the company struggled to find a consumer market for Fiberglas. Draperies and insulation could only take the company so far. But it was the discovery of the ability to reinforce plastic resins by using Fiberglas that set the company on its profitable path.

Fiberglas-reinforced plastic had been used in some military applications where strength without weight was required. After the war, Owens-Corning attempted to market it to the consumer by producing gimmicky products like Fiberglas fishing rods and golf clubs. But by 1952, Fiberglas-reinforced plastic found a market in applications to bodies of boats and automobiles. General Motors featured it in its newest sports car, the Corvette, in 1953.[37] The product allowed sleek, molded bodies without the weight of steel, hence allowing the car to go faster. Owens-Corning helped underwrite a new plant in Ashtabula, Ohio, to produce the Corvette bodies. Fiberglas-reinforced plastics pushed Owens-Corning's sales pass the $100 million mark that year.

The company also continued to push its original product lines, including insulation. Research-and-development efforts improved the insulation substantially, and the company sought ways to increase its use. In 1957, Owens-Corning announced the "Comfort Conditioned Home Program," described as "the most comprehensive marketing campaign in Fiberglas history and an outstanding promotion in the interest of the home building industry."[38] With the cooperation of utility companies, Owens-Corning designed a house that could be heated in the winter and cooled in the summer for just $10 per month. It was the brainchild of employee Tyler Stewart Rogers, who urged builders to construct homes with three inches of insulation between wall studs—twice as much as was generally used in new home construction—to improve comfort and reduce energy costs. Builders were also encouraged to utilize other Owens-Corning products as optimally as possible. Advertisements showed houses with Fiberglas roofing, ductwork, draperies, acoustical tiles, reinforced wallboard, and furnace filters. The program was helped through recognition by the Federal Housing Administration's home loan program, which encouraged new homeowners to buy better-insulated homes. In 1958, 625 home builders participated in the program, building 40,000 houses insulated with Owens-Corning Fiberglas. The "Comfort Conditioned Home Program" made possible the expansion of housing into geographic regions in the South and Southwest previously ruled out for development because of the heat of the summer, such as Albuquerque, New Mexico.

Research, development, and promotion of new products continued in the early 1960s. A new technical center opened in Granville, Ohio, in 1960, employing more than 250 scientists and engineers. As it had done at the New York World's Fair of 1939, Owens-Corning also displayed its products at the 1964 New York fair. Also that year, the company reported on experiments to produce Fiberglas-reinforced rubber tires. Contracts with the military and aerospace industries opened up new markets that would expand greatly by the end of the decade.

Two major personnel changes occurred in the company in the middle of the 1960s decade. In 1963, Dr. Games Slayter, who was one of the creators of Fiberglas in the early years of the company and called the "Father of Fiberglas," who held more than 160 patents and who had been a vice president of the company, retired. He died suddenly less than a year later.

Also in 1963, General Lauris Norstad, the former supreme allied commander, Europe and commander in chief of the U.S. European Command, was elected president of Owens-Corning, and Harold Boeschenstein moved into the position of chairman. Boeschenstein, who had headed the company since its founding, retired completely in 1967, and General Norstad became chairman and chief executive officer with John H. Thomas serving as president. Norstad, in assuming the position of leader of the company, paid tribute to Harold Boeschenstein. "Taking over as chairman of Owens-Corning Fiberglas, I am mindful of the responsibilities as well as the honor involved in leading this great company. I am equally mindful of the fact that no one can quite fill the shoes of Mr. Boeschenstein who founded and built the business and who has earned world acclaim as an industrial leader. Our task is to continue our company's growth and, if we all dedicate ourselves to achieving our hopes and aims for Owens-Corning, together we will succeed."[39] The company was not without a Boeschenstein in its leadership, however. Harold's son William was appointed executive vice president, a move up from his position as vice president for marketing.

To the Moon, the Arctic, and the Desert

Its experience during World War II showed Owens-Corning that its products had applications to the aerospace and defense industries. In 1962, the company created the Aerospace Division to handle contacts (and hopefully contracts) with the government. Early applications of Fiberglas to defense included converting Fiberglas boats into shallow-draft river patrol boats for the Vietnam War.[40] The company also built components for ballistic missiles, body armor for soldiers, and heat shields. The strength, light weight, and heat resistance of Fiberglas made it a natural for many defense applications.

But it was tragedy that led to the most notable application of Fiberglas in the space industry. In 1967, the United States inaugurated its Apollo space mission with the ultimate goal of landing a man on the moon. Apollo 1 was to be the first manned test flight of the Command/Service Module that would be used to get to the moon by future Apollo flight crews. It was the most complex spacecraft ever developed by NASA, and the three-man crew—Virgil Grissom, Edward White, and Roger Chaffee—had expressed concerns to superiors about the amount of flammable materials in the spacecraft cabin. In a launch simulation test on January 27, 1967, the spacecraft, filled with 100 percent oxygen, caught fire and within seconds flames engulfed the entire cabin. The astronauts were unable to open the hatch to escape, and all three died.

Within weeks, Congress began a complete investigation of the tragedy. At a hearing by the Senate Committee on Aeronautical and Space Sciences held one month after the fire, investigators noted the need to replace flammable cloth with nonflammable materials. This included the fabric that was used to make the astronauts' space suits. Testimony before the Senate committee suggested using Beta cloth, a material that could withstand up to 1,500 degrees without burning.[41] Beta cloth had been developed by Owens-Corning in 1962, and was originally used to make products like draperies and bedspreads. It later was improved by coating the threads with Teflon.

Beta cloth made its debut on the Apollo 7 flight that orbited the moon. It was used not only in the space suits but also in sleeping bags and as equipment covers used in the flight. The company touted it for its fire resistance properties and its ability to insulate the astronauts who would eventually land on the moon from temperature extremes.

With the successful moon landing of Apollo 11 in July 1969, Owens-Corning began an all-out public relations campaign focused on the Beta cloth space suits that were used on the moon. One of the space suits was displayed in the window of the Fiberglas Center, a product exhibition space on Fifth Avenue in New York. The company produced press kits that included mission designation patches like those worn by the Apollo 11 astronauts. But all of the publicity caused the company some problems. In August 1969, Owens-Corning received a letter from lawyers representing ILC Industries, Inc., the company that actually manufactured the space suits, expressing its concern that Owens-Corning was promoting itself as the manufacturer of those suits. The display on Fifth Avenue, for example, urged the public to come in to "see our space suits." ILC's lawyers stated their displeasure bluntly. "Your company's efforts to curry public favor by taking far more credit than is its due for its part in the Apollo project, is evident."[42] The letter noted that Beta cloth was just one of the 21 layers of the spacesuits. Owens-Corning also had to deal with complaints from NASA that the cloth was not durable enough, and that it itched when in contact with bare skin. Owens-Corning, in turn, expressed its dismay to NASA that while the astronauts' suits were made of fire-safe fabric, the flags taken to the moon and planted on its surface were not.[43]

In October 1972, Harold Boeschenstein died of a heart attack in New York City. Boeschenstein was remembered by many for his success in taking Owens-Corning Fiberglas from $3 million in sales in 1939 to nearly $300 million in 1964.[44] Business leaders from around the country came to his funeral service, including Henry Ford II of Ford Motors, Edgar Kaiser of Kaiser Industries, and Amory Houghton of Corning Glass. Toledo business leaders such as Ward Canaday and fellow glass industry leaders including Robert Wingerter praised Boeschenstein's leadership. Otto Wittman, director of the Toledo Museum of Art, recalled Boeschenstein's generosity. The year before his death, Harold's son William took over as president of the company, succeeding John H. Thomas.

Under William Boeschenstein, the company continued it growth, with new products developed by its engineers. These products included a one-piece tub and shower unit that was used in many new hotels, including the Renaissance Center built in 1976 in downtown Detroit. Teflon-coated Fiberglas fabrics were used to cover several sports stadiums, including the Pontiac Silverdome. One of the largest projects undertaken by the company was insulating the trans-Alaskan pipeline—some 400 miles of it that stretched from the state's North Slope over the mountains to the port of Valdez.

Creating the insulation for the pipeline was a process that took five years to perfect. It required designing new manufacturing equipment to make the 70,000 sections and 35,000 support modules. At its time, the trans-Alaskan pipeline was the largest insulation job in history.[45] Construction of the pipeline began in 1974 after the Arab oil embargo threatened the United States energy supply, and it was completed in 1977.

In 1978, Owens-Corning moved from the arctic to the desert with another massive engineering project—a new airport terminal at the King Abdul Aziz International Airport near Mecca in Saudi Arabia. The terminal was an effort by the Saudi government to facilitate the more than 1.5 million people who visited the holy site each year in the pilgrimage known as the Hajj. The new terminal had to be able to shield 50,000 to 80,000 air travelers at a time from the 130-degree heat typical for the desert. It also had to in some way reflect the history and culture of the region.[46] The solution was an expansive fabric structure that appeared like a series of tents. While several large fabric structures had been constructed in the early 1970s, the Hajj airport terminal was the largest. It was designed by Skidmore, Owings, & Merrill, which awarded the contract for the fabric to Owens-Corning in 1978 based on the company's extensive experience in designing and erecting fabric roofs.

Because of the massive size of the project, Owens-Corning created a prototype of the roofing system at the company's technical center in Granville, Ohio. The materials had to be produced in prefabricated units because each section was built in the United States and then shipped to Saudi Arabia for construction. With only extremely small tolerances allowed, the company relied on sophisticated computer models to help guide production. The prototype alone cost $2.5 million.[47] But what was learned from the prototype made construction in the desert run smoothly. The tent roof was supported by 440 150-foot steel pylons that were covered by 5.5 million square feet of Fiberglas Beta fabric coated with Teflon. The fabric was held in place by high-tension cables that were, in turn, held together by a ring at the top. Each section was raised like a tent. In a publication about the project, Owens-Corning described it as "a marvelous melding of ancient and modern. Islamic tradition and the centuries-old use of tents are unified and symbolized by the power of today's engineering."[48]

The Hajj terminal received numerous awards for its design. In 2010, the American Institute of Architects awarded it the organization's 25 Year Award in recognition of its enduring significance to architecture.

At the end of the 1970s decade, Owens-Corning Fiberglas under the younger Boeschenstein's leadership was prepared to continue its expansion. William Boeschenstein explained at the annual stockholders' meeting in 1979 the company's four tenets for future growth that were to guide the company, not unlike the "Four Basics" that had guided the company from its founding by his father in 1938.[49] These included the expansion of markets for traditional, core products like insulation and textiles; expansion into businesses that were well established, but compatible with Owens-Corning; establishing new businesses that could exploit the uniqueness of Fiberglas; and select expansion of international business ventures. To help achieve these new thrusts of development, the company built the largest glass fiber reinforcement plant in the world in Amarillo, Texas, which opened in June 1979. The state-of-the-art plant was completely computerized, and also featured an "employee total work environment" philosophy.[50] Workers were encouraged to communicate with each other and with management to develop a sense of teamwork and improve productivity. They worked three 12-hour shifts in order to have more free time. The interior of the factory was painted with stimulating colors and decorated with wall-sized murals.

In 1980, advertising executives for Owens-Corning approached Boeschenstein about a possible new "spokesperson" for the company. As Boeschenstein later recalled, "I was expecting Jack Nicklaus or Arnold Palmer or Farah Fawcett or somebody that was important."[51] He was surprised when they proposed the Pink Panther, a cartoon figure who did not even speak. "And it became apparent to the group very shortly into their presentation that I didn't have a clue as to who the hell the Pink Panther was," he stated. "But I didn't want to embarrass them too much. So I said, okay you go out and do a survey of various age groups, you know, that people up to 30, 30 to 45, 45 to 60, and over 60. Well the survey came back, and everybody knew the Pink Panther, including people over 60 or 65. Everybody, I mean it was 95, 98 percent. And so I still wasn't sure about it, so I asked my mother who was then about 90 whether she knew who the Pink Panther was and she said yes. So that was it. We went with the Pink Panther," Boeschenstein recalled. The character reinforced the tie between the color pink and the company's insulation color. Consumers may not have recognized the technical advantages of Owens-Corning insulation over other brands, but they knew it was pink. The campaign was extremely successful for the company, and it continues today.

Owens-Illinois: Timber, Sugar Cane, and Paper Plates

Like its counterparts Owens-Corning Fiberglas and Libbey-Owens-Ford, Owens-Illinois also expanded and diversified in the decades after World War II. The company also expanded globally, which caused it some difficulties when it was swept up by world events.

In 1950, the reigns of the company were handed over from William Levis to his cousin J. Preston Levis, who became chairman and chief executive officer. Carl R. Megowen was elected president. Like Libbey-Owens-Ford and Owens-Corning, Owens-Illinois invested heavily in research and development in the 1950s, opening a new $10 million technical center in 1955 that provided facilities for the complete study of glass, from theoretical analysis to consumer testing. The facility housed 400 engineers, scientists, and technicians in 200,000 square feet of space.[52] The company hoped that, with all research-and-development efforts in one place, new products could be developed faster and more efficiently, and duplication in research could be avoided.

The company's international expansion included opening new factories in Cuba, Panama, and Venezuela in 1956. The Havana, Cuba, plant was the only glass container plant in that country. Unfortunately for Owens-Illinois, it was nationalized in 1960 by Fidel Castro after his revolution, and O-I's investment was completely lost.[53] The Venezuelan operation, which made glass containers and window glass, would also become a source of trouble for the company.

In 1956, the company continued its real estate expansion, both foreign and domestic, when it acquired the National Container Company. National Container operated six paper mills, 18 box plants, and two bag plants and owned nearly 1 million acres of timberland, some of it in places such as the Bahamas.[54] For Owens-Illinois, having a company that could produce shipping boxes for its many products made sense. National Container would become O-I Forest Products Division in the 1960s.

The Bahamas timber operation consisted of 300,000 acres leased from the British on Great Abaco Island, about 75 sea miles from Grand Bahamas Island. Because of its remoteness, Owens-Illinois had to build roads, power-generating plants, and worker housing. It converted a 50-year-old cruise ship, the SS *Robert Fulton*, into employee apartments and anchored it off the island.[55] Large equipment was brought in to build roads through the dense forest, with routes mapped out by an amphibian airplane that flew over the island. Crews would fan out from the roads, cutting timber with chain saws and hauling out the cut logs with tractors. The crews cut about 2,000 cords of wood each week that was pulped at a paper mill on the island. In order to more effectively run its forest product division, in 1960 Owens-Illinois built a laboratory dedicated to research on timber near its technical center on Westwood Avenue in Toledo.

In 1966, Owens-Illinois moved in a entirely new direction on the Great Abaco Island. In order to ease the economic impact of ending its timber-harvesting business on the island's population, it decided to grow sugarcane. The climate and the improvements already in place on the island for the timber business convinced the company that this was a good decision. The operation also took advantage of a new sugar quota system approved by the U.S. government in 1965 that sought to ensure a low-cost supply of sugar from neighboring friendly countries. The company believed the quota of refined sugar that would be allowed to be imported from the Bahamas would be at least 10,000 tons.[56] It also estimated that the sugarcane operations would produce a profit margin of 12.8 percent, and the initial costs would be recovered in six years.[57] The Bahamian government leased the company 21,000 acres of land at no cost for the operation, with an option for 25,000 additional acres.

An analysis conducted by consultants to Owens-Illinois optimistically reported that "large sugar operations conducted by private enterprise have proven attractive and profitable investments."[58] Owens-Illinois executives also felt that Great Abaco could eventually develop as a recreational and residential island, thus ensuring long-term profitability. Unfortunately, the sugarcane operations proved anything but profitable. The 1969 harvest yielded only 15,000 tons of raw sugar, and the 1970 harvest only 20,000 tons, neither of which produced enough refined sugar to provide a margin for profit. In 1970, the company took a $22 million charge off its profits in order to discontinue the sugar operations.[59]

Despite the failure of the sugar plantation, the Forest Products Division of Owens-Illinois became the second largest unit in the company. In addition to corrugated cardboard shipping containers, the division also produced plastic, dimensional lumber, and composite cans that consisted of cardboard tubes with metals ends.

In 1962, Owens-Illinois suffered the loss of William Levis, who had served the company for nearly 50 years. In a tribute by the company's board of directors, Levis was described as "the architect who blueprinted the directions and dimensions of growth and made Owens-Illinois a successful industrial enterprise."[60] That year also saw the loss of Carl R. Megowen, who died unexpectedly at the age of 63; and Herman K. Kimble, who had worked for the Kimble Glass Company since 1917. In 1963, J. Preston Levis became chairman of the board of directors, and Raymon Mulford took over as president.

Under Mulford, the company continued both its physical expansion and its diversification. In 1966, the company purchased almost 400 acres of land in Perrysburg, Ohio, to be used as a manufacturing site for pilot projects and new product development. The site was named the Levis Development Park, in tribute to both William Levis and J. Preston Levis. The park was intended for smaller-volume projects that could help bring new products to full commercial development.

Mulford held a progressive view of corporate leadership. He was committed to not only making a profit, but also benefiting society through the actions of his company. He was concerned about worker safety, about the environment, and about diversifying the company's workforce. In a book chapter published by the American Management Association in November 1966, Mulford addressed the state of workforce integration at Owens-Illinois. In 1962, Owens-Illinois signed a "Plan for Progress" pledge along with many other U.S. companies, promising to work to recruit more African Americans and ending discriminatory hiring practices. Mulford said the company had experienced difficulties in finding African American workers for technical and management positions, but that the company was committed to doing so. "From a purely pragmatic point of view, it makes good sense for business to promote job equality. It is in our interest to help bring minority groups into the mainstream of our economic life—as workers and consumers—so they can be contributors and participants, rather than recipients of welfare and cause higher taxes," Mulford wrote.[61]

In 1968, Mulford became chairman of the board of directors of Owens-Illinois, and Edwin D. Dodd assumed the position of president from general manager of the Forest Products Division as J. Preston Levis retired. Also that year the company made another major diversification purchase by buying Lily-Tulip Corporation, which made paper plates and cups for both the consumer and commercial markets. Mulford and Dodd stated that they felt the merger was a good one, moving the company into convenience products that had the potential for growth. Mulford died unexpectedly in 1973, and Dodd became president.

Environmental Concerns

Because much of its product line consisted of disposable containers, the environmental movement had a direct impact on Owens-Illinois. After all, it was O-I that had pioneered the concept of a one-way glass bottle in the 1940s. What was once seen as convenient and progressive came to be seen as wasteful and polluting.

To combat bad publicity related to the public's antilitter concerns, Owens-Illinois was one of the founding members of Keep America Beautiful, Inc. in 1953.[62] Smith L. Rairdon led the company's involvement in the organization. Other companies that belonged included Philip Morris, Anheuser-Busch, and Coca-Cola. In reaction to the growing litter problem along interstate highways, Keep America Beautiful developed public service advertising aimed at reducing "litterbugs." The Lily-Tulip division also began to promote antilitter messages printed on its disposable paper cups.

But Owens-Illinois fought efforts to ban nonreturnable containers and to tax them by requiring deposits, arguing that disposable containers were a reflection of an affluent society that demanded convenience. The company was supported in its stance by William Ruckelshaus, director of the U.S. Environmental Protection Agency, who stated that the government would not pass laws restricting the production of nonreturnable bottles and cans.[63] Owens-Illinois established the first glass recycling program in the nation at its Bridgeton, New Jersey, plant in 1968. In 1970, the program expanded to other plants, collecting 29 million glass containers that year that were melted down to make new ones.[64] By 1977, it passed the one billion pound mark in glass recycling.[65] The company also supported other efforts to develop a modern solid waste collection system, including a nationwide recycling program.

The company's Forest Products Division was another source of pollution. Concerns included emissions into the air and contamination of water at its paper mills.

While Owens-Illinois was also one of the first companies in the United States to adopt an official corporate policy on the environment, it was also concerned about the cost of environmental controls, particularly regulations that made only small improvements in environmental performance at substantial cost.[66] "Excessive cost increases to meet debatable control levels can deter purchases, reduce production and employment, and have a detrimental effect throughout the entire economy," the company noted in its 1974 annual report.[67]

The Energy Crisis: A Blessing and a Curse

While the United States had periodically experienced shortages of energy sources throughout its history, the crisis of 1973 had a lasting impact on the nation. In October of that year, the Organization of Petroleum Exporting Countries placed an embargo on the export of oil to the United States by its member countries. The embargo was in response to the United States backing Israel in an armed conflict with Egypt and Syria. While the military conflict was the initial reason given for the embargo, OPEC also saw the need to have greater control over the supply of oil in order to stabilize prices and protect profits. So while the embargo ended in March 1974, its effects continue today.

Both the general public and industry soon recognized that unlimited and cheap supplies of energy were a thing of the past. Energy conservation became not just a socially responsible concept for industry, but a must because of the spikes in costs for all forms of energy created by the crisis. There was also a new focus on domestic energy supplies, resulting in the approval of the long-stalled trans-Alaskan oil pipeline in 1974.

The need to control energy costs became evident again in 1979 when a second energy crisis hit the country as a result of the Iranian Revolution that nationalized that country's oil production. After American hostages were seized by the Iranian revolutionaries, President Jimmy Carter stopped importing Iranian oil. Once again, long lines appeared at gas stations. This crisis did not end as quickly of the first energy crisis because with the outbreak of war between Iran and Iraq in 1980, nearly all oil production in both countries ended. As in 1973, the crisis in oil led to shortages in other energy sources like natural gas as industries converted to other energy sources. Environmental regulations that restricted how coal could be burned further complicated the energy picture.

Another factor in the energy equation for glass companies was inflation resulting from the economic recession of the late 1970s. Interest rates on borrowed money increased substantially, which made product diversification more expensive for the glass companies seeking to reduce their glass production and modernize their facilities to be more energy efficient.

How Toledo's glass companies were impacted by the energy crisis of the 1970s depended on what products they produced and whether their products could be seen as part of the energy solution or part of the problem. For a company like Owens-Illinois, which produced mostly products that consumed energy, the crisis meant higher prices and the expensive need to convert factories to alternative energy to operate its glass furnaces. Owens-Illinois did promote its glass-recycling program as a way to save energy and protect the environment by reducing the need for raw materials.

One energy-saving product developed by Owens-Illinois was the SUNPAK solar collector.[68] It used the company's experience with glass tube production to create a system that could collect the sun's energy and use it to heat or cool homes. The company concentrated its development on commercial applications. By 1979, it had installed 57 systems.

For Libbey-Owens-Ford, the energy crisis impacted its bottom line. The automobile industry felt the crisis as people stopped buying large, gas-guzzling cars. As the number of cars purchased declined, the need for windshields for those cars also declined. In addition, the smaller cars that the auto industry began to produce required smaller, less profitable windshields. One growth market was windshield replacements as consumers kept cars longer.

As energy costs increased and markets for L-O-F products decreased, the company sought to reduce labor costs. Layoffs even impacted salaried workers, whose numbers were reduced by 10 percent in 1974 and 1975.[69] Some plants were closed (such as Shreveport, Louisiana), and others were converted to float glass production to hold down labor costs.[70] The company also changed its energy source when necessary, especially when natural gas supplies were not available. In 1976, the company contracted to drill its own natural gas wells in southeastern Ohio to stabilize its supply. The wells cost $2 million to drill, and the company spent another $6 million on additional wells.[71] The effort was necessary when natural gas supplies in Toledo were curtailed.

There were two L-O-F product lines that were able to be marketed to the post-energy crisis consumer—insulating windows and solar energy panels. The company had always marketed its Thermopane windows as energy efficient. The company improved the efficiency by developing new coatings for the glass called Vari-Tran.

Like Owens-Illinois, Libbey-Owens-Ford also entered the solar energy business. In 1974, the company began to research thin film coatings for solar collectors in a pilot project. In 1977, the company acquired 39 percent interest in Photon Power, Inc. of El Paso, Texas, which was jointly owned by a French company.[72] It did research on photovoltaic cells that would allow direct conversion from solar energy to electricity. The company also came out with its own solar collection product called SunPanel, which was intended for both commercial and residential use. Prototypes of the collectors were installed at the University of Toledo, and the company partnered with UT researchers to study their effectiveness. SunPanel was selected for installation at several large public institutions in 1976.[73] As the company's annual report for that year stated, "Although a general transition to solar heating will proceed slowly, the outlook for future growth appears favorable."[74]

While Libbey-Owens-Ford had some limited success in turning the energy crisis into a plus for the company, the most successful Toledo glass company in energy-related products was Owens-Corning Fiberglas. The company's long history in insulation production positioned it to make major inroads into the consumer energy-saving market. In 1974, the company put on an exhibit at its Fiberglas Exhibit Center on Fifth Avenue in New York City called "The Energy Show."[75] It featured live demonstrations on how the average consumer could save energy.

The success of the company in marketing its products as energy-saving ones can be seen in its sales figures. In 1974, just as the first energy crisis was hitting home, the company recorded $800 million in sales. By 1978, that figure had risen to $1.7 billion, at a time when an economic recession was negatively impacting most other corporations.[76] William Boeschenstein continued to expand the company, spending $200 million that year to increase production capacity. The company not only marketed its insulation for its potential energy saving, but also products like shingles and bath and shower enclosures.

Despite the difficult times experienced by the auto industry, Owens-Corning also saw the energy crisis as a way to expand its inroads into automobile production. A story in the *New York Times* in June 1978 titled "Owens-Corning Soars on the Fiberglass Boom" noted that while Owens-Corning was the largest supplier of insulation, it was counting on expansion in other areas of production like Fiberglas-reinforced plastics to continue to boost sales.[77] The company was particularly betting on increased used of the composites in automobiles, not only to lighten them, but to stop corrosion, a major factor in automobile aging.

Storms on the Horizon

Toledo's glass companies entered the decade of the 1980s with a mixed picture for the future. The diversification that had greatly expanded their operations had come at a huge cost as interest rates escalated during the inflationary recession of the late 1970s. As Edwin Dodd of Owens-Illinois predicted in the company's annual report of 1974, "As we look ahead, our four most important challenges are inflation, recession and increasing unemployment, high long-term interest rates, and the shortage of energy."[78]

But that did not stop the diversification efforts. In 1981, Owens-Illinois expanded into an unusual area, given its history—health care. Based on the fact that several of its divisions such as Kimble and Libbey were supplying products to the health care field, the company acquired a minority interest in Health Group Inc., a Nashville, Tennessee-based hospital management company, for $10 million (which it renamed Health Care and Retirement Corporation of America.)[79]

While Owens-Illinois's international subsidiaries had helped it to expand, this expansion had come at a price for the company. Its Cuban operations had been nationalized in 1960 during the Cuban revolution. In 1976, William F. Niehous, O-I's top executive in Venezuela, was drugged, kidnapped, and taken from his home in Caracas by six Communist sympathizers hoping to foment revolution.[80] His wife, who was bound and gagged by the guerillas, was able to escape. Held for three years in the jungle in chains, Niehous

The Kaylo pipe insulation product that would cause many problems for workers and for its maker, Owens-Corning Fiberglas. (Ward M. Canaday Center for Special Collections. Used by permission of Owens Corning.)

dramatically escaped in June 1979, making headlines around the world. Niehous returned to Toledo, where he became vice president and director of administration for corporate technology at the company's headquarters. His kidnappers were arrested and served time in prison. But the episode pointed out again the threats that world events could have on corporate operations.

In 1981, Owens-Illinois sold its Lily-Tulip Division to the investment firm Kolhberg Kravis Roberts & Co. for $150 million. The money raised from the sale was used to pay off some of Owens-Illinois's debt. This would not be the last of O-I's interactions with KKR.

At Owens-Corning Fiberglas, the rosy predictions of William Boeschenstein for continued growth were suddenly dashed in 1981 when the company was sued over the production of Kaylo insulation. Kaylo was originally manufactured by Owens-Illinois, which began selling it in 1943. The insulation was mostly used around pipes such as those in furnaces and other high-temperature applications because of its heat-resistant qualities, which were the result of the inclusion of asbestos in the product.[81] In 1958, Owens-Corning, which had already been acting as a distributor of Kaylo, purchased the product line from Owens-Illinois along with the factory in Berlin, New Jersey, where it was produced.

Owens-Corning's 1981 annual report noted that it was the subject of nearly 12,000 lawsuits resulting from employee and consumer claims of mesothelioma, a fatal cancer caused by exposure to asbestos fibers.[82] While the company had stopped making Kaylo that contained asbestos in 1972, it had sold $135 million worth of the product since 1958. The number of lawsuits filed against the company would grow exponentially in the early 1980s, and would require a dramatic solution.

EIGHT

Like Being Nowhere at All

Saturday night in Toledo, Ohio,
Is like being nowhere at all.
All through the day, how the hours rush by,
You can sit in the park, and you watch the grass die.

—Lyrics to "Saturday Night in Toledo, Ohio,"
by Randy Sparks, 1969

On June 11, 1973, a rising star in popular music named John Denver appeared for the first time on *The Tonight Show Starring Johnny Carson*. In his debut performance on this highly rated late-night show, Denver chose a song written by Randy Sparks, formerly of the New Christy Minstrels, that Denver often sang in concert, but had never recorded. The performance of that song on national television—entitled "Saturday Night in Toledo, Ohio"—forever changed the subject of its stinging satire.

City fathers took notice. The decay of Toledo's downtown had distressed government and business leaders for decades, and various plans had been hatched over the years to address the embarrassment. Yet there was little to show for all of the planning. For many, the derogatory lyrics rang true.

While government and business leaders were concerned about the image of the downtown, Toledo voters were not. Denver's performance came less than three months after the stunning defeat by 64 percent of Toledo voters of a plan to spur downtown redevelopment by building a convention center. The song sparked a new effort at downtown revitalization in spite of this rebuke by the voters. Within days, a new organization of business leaders (including representatives of the powerful glass industry) was formed to lead new—and hopefully successful—downtown redevelopment activities. The goal was to turn the decaying downtown into a world-class city, and plans evolved to center on a sparkling new headquarters for Owens-Illinois. It would also include an international hotel, a new headquarters for the Toledo Trust bank, a county-city-state office building, and a "festival marketplace," among other efforts. Today, many of these grand schemes have either changed dramatically or failed miserably, and the "Saturday Night in Toledo, Ohio" song's lyrics still haunt the city. The story of how downtown redevelopment unfolded reveals much about the interplay of corporate leaders (particularly those at Owens-Illinois), elected officials, the media, and the citizens of the city at an important juncture in Toledo's modern history.

Early Redevelopment Efforts

The downtown redevelopment efforts in the 1970s and early 1980s were the realization of the dream begun 40 years before with the "Toledo Tomorrow" exhibit. "Toledo Tomorrow" clearly sparked the imagination of the city, and almost immediately after the exhibit closed numerous groups began planning for a real "Toledo Tomorrow." In 1945, the Metropolitan Planning Committee of the Toledo Chamber of Commerce issued its master plan for the city. In 1951, Harold Bartholomew and Associates produced a plan called the "Toledo Urban Area" report. In 1957, the Toledo–Lucas County Plan Commission produced "The Hub of Toledo," which focused on the downtown and called itself a "plan for tomorrow." None of these produced much actual development. The new headquarters building of Libbey-Owens-Ford in 1960 and the Riverview development project of 1969 (including the Fiberglas Tower) were the two most significant outcomes of the postwar downtown redevelopment efforts.

In 1969, Owens-Illinois, one of the largest employers in the city and one of the major tenants of downtown office space, undertook its own long-range planning aimed at examining its space requirements until the year 2002 and its prospects for staying in the downtown.[1] During the discussions that produced these Owens-Illinois facility reports, the company's executives, including Raymon Mulford and Edwin Dodd, began talking with others interested in downtown redevelopment, including Harold Boeschenstein of Owens-Corning, Stephen Stranahan of Entelco, Charles McKelvey of the First National Bank of Toledo, developer Dean Bailey, and Toledo mayor Harry Kessler. The Urban Land Institute, a nonprofit city-planning corporation, approached Stranahan and proposed that it assist with downtown planning. With the approval of the mayor, Owens-Illinois, Libbey-Owens-Ford, Owens-Corning Fiberglas, and Entelco agreed to sponsor the study, and ULI began work on the project in late 1972.[2] ULI was to study the downtown area bounded by the river, develop a program of action, and test the feasibility for the plan's success. Unfortunately, Raymon Mulford's involvement with the project was cut short by his death from a heart attack in February 1973.

As ULI was undertaking its study, Toledo's unemployment rate was rising and employment patterns were shifting from the high-paying manufacturing sector to lower-paying service sector jobs.[3] While the population of the city was still growing, that growth was the result of 20 years of annexation, not from new migration. The city was facing a major budget deficit, and city services were being cut.

Almost simultaneously to the Urban Land Institute's planning efforts, a group of business and political leaders formed the Convention Center Committee in April 1969 in an attempt to revitalize the downtown by building a convention center and auditorium. A pivotal participant in the plan was the University of Toledo, which was looking to replace its aging basketball arena and build a new continuing education center. The Convention Center Committee wanted to include one or both of these elements in its plans because the committee believed that UT's participation would be vital to the convention center's success. The chair of the Convention Center Committee was Charles McKelvey, president of the First National Bank of Toledo. In 1971, McKelvey resigned from the committee to become a member of the UT board of trustees, a position where he could potentially influence the university's decision on whether to participate in the project.

But plans for the convention center dragged on. In early 1973, a citizen's group called ALERTA (which stood for Affiliated Leagues for Equal Representation and Taxation Alliance) circulated petitions to get the convention center issue placed before Toledo voters, and succeeded by collecting 20,000 signatures. The group tapped into the resentment of many who failed to see how the city could go forward with a downtown convention center when basic city services were being cut to solve a budget deficit of $11.7 million. There was also growing resistance to the idea of building UT's basketball arena away from campus.[4] Proponents of the convention center, called the Committee for Progress, countered it would create over 3,000 jobs, mostly for those of lower incomes; would produce over a $1 million in new revenue; would benefit the entire community by generating $38.6 million in economic gains; and would be paid for by revenues from a hotel-motel tax that had been designated for the project.[5]

On March 20, 1973, the effort to build the convention center was dealt a fatal blow when a charter amendment that made proceeding with the project all but impossible passed with 64 percent of the vote. ALERTA hailed the vote as a step toward reordering the city's capital expenditure priorities.[6] The Toledo *Blade* countered in an editorial the day after the vote, "While the convention center is itself a desirable facility, and may be delayed indefinitely, it would be regrettable if the results yesterday were to be interpreted as an unwillingness by the city to make any substantial commitment to the downtown area at the very time it is looking to the private sector for major help in enhancing the heart of the city."[7]

It was against this backdrop that the Urban Land Institute issued its report in early April 1973, less than three weeks after the vote.[8] While recognizing the glum mood of the city, the ULI report did find some reason for optimism. First and foremost was what ULI saw as the sound economic underpinnings of the city that included seven Fortune 500 companies, including its three glass companies: Owens-Illinois (ranked 77th), Dana Corporation (199th), Owens-Corning Fiberglas (224th), Libbey-Owens-Ford (225th), Champion Spark Plug (339th), Questor Corporation (373rd), and Sheller-Globe (458th). There was also a retail sector in the downtown with three major department stores. The ULI report also noted the city's greatest asset—its riverfront—which remained undeveloped or dominated by heavy industry. Of particular interest to ULI was the Middlegrounds area, which at the time belonged to three railroad companies and housed some derelict terminal buildings. The location of the Middlegrounds along the riverfront made it a prime space for residential development, the ULI believed.

But ULI also found the downtown lacking in several important ways: there were no significant cultural attractions; the Sports Arena, where hockey was played and concerts held, was aging and inadequate; there was no first-class hotel downtown; and the central business district was too spread out, which made it hard for visitors to move from one activity to another. The downtown lacked focus.

The ULI also noted the most important barrier to downtown redevelopment that had stopped other efforts from succeeding: the will of the people. ULI said the city's population believed there were other priorities for the city than developing the downtown. "There seems to be general unawareness of the fact that neither Toledo nor any other major metropolitan area will have a healthy body politic if the heart it surrounds is seriously deteriorating or is denied corrective action," the plan warned.[9] All of the previous plans touted for redeveloping the downtown had been made by individuals with no authority to implement them, the planners found, and the citizens had no expectation that any would be executed. "We suggest that both the business community and the forces of government need to keep in better touch with the electorate and citizens of metropolitan Toledo, who are in many ways the constituents of both," the report stated.

Lastly, the ULI report made a bold assertion: despite the vote of three weeks before, the report suggested the city proceed with a convention, exhibition, and sports complex. The ULI panel stated that other cities of similar size had found convention centers to be valuable in bringing life to the downtown, especially in the evening hours.[10]

Two months after the ULI report was released, and three months after the defeat by the voters of the convention center plan, John Denver made his infamous appearance on the *Tonight Show*. The plight of downtown became an embarrassment to Toledo before a national audience. Randy Sparks, in an interview with *Folk Alley Extras* years later, said his inspiration for the song came during a one-night visit to Toledo while touring with his band, the New Christy Minstrels, in late 1960s. "When we got there, it was closed," Sparks said. Sparks took credit for inspiring Toledo to revitalize the downtown: "Eventually they had meetings and they decided they were going to fix the town because of my stupid song," he recalled.[11]

Toledo Looks to the River

Within five days of Denver's *Tonight Show* performance, the *Blade* reported on the creation of an organization named the Greater Toledo Corporation (GTC) to promote redevelopment of the downtown area as outlined in the Urban Land Institute report. Edwin Dodd was named chairman of the committee, signaling the leadership role that Owens-Illinois would assume in downtown redevelopment. As recommended by the ULI, the GTC and the mayor created a second public group, the Toledo Development Corporation (TDC) in April 1974, to ensure sufficient public leadership on any future redevelopment. The GTC and the TDC were the two links that the Urban Land Institute suggested to produce consensus between the public and private sectors on redevelopment, hopefully avoiding the disaster that happened with the convention center when private business interests were out in front of the public's desires.

Mayor Kessler charged the TDC to prepare a development program to enable the downtown to capitalize on opportunities; to serve as a policymaking and advisory group on revitalization projects; to coordinate between citizen, business, professional, and civic leadership; and finally, to set forth a manageable future for downtown Toledo. The TDC was to work with the GTC to produce an achievable master plan that would serve as the catalyst for the entire downtown revitalization effort.

The TDC had 11 members, all of whom were appointed by Mayor Kessler. The group included only one person from outside the business community—Betty Mauk, community activist and promoter of a small green space on the riverfront called Promenade Park. But because of criticism that this group was not broadly representative, a Citizens Development Forum, made up of 61 community members, was established to serve as an advisory board to the TDC. A Technical Advisory Committee was also created, which consisted of nonelected government leaders who were charged with carrying out the long-range plans of the TDC. And lastly, there was the GTC, the nonprofit organization with representatives from 15 of the city's corporations, including the glass companies. Leslie Barr, the head of a Louisville, Kentucky, planning firm, was hired by the GTC as its president in December 1973. To further complicate matters, three consulting firms were hired. With so many groups involved in downtown revitalization, the potential for conflict existed from the start.

For the first year, the consultants hired by the GTC and the TDC worked on producing their reports built upon the work of the Urban Land Institute. In July 1974, the consultants issued a preliminary report that suggested three to four new office buildings, including new headquarters for Owens-Illinois; an additional office building for Owens-Corning Fiberglas; a new city-state office building to replace the aging Safety Building, to be located at the corner of Adams and Summit; and a galleria shopping area that would connect the Lion's and LaSalle's department stores.[12]

The final report of the consultants, which focused on riverfront development, was issued on May 20, 1975. Entitled "Toledo Looks to the River," the report was quickly endorsed by the TDC and the Toledo Citizens Forum.[13] The 45-page color book critiqued the current state of the Maumee River in the downtown area, which the authors found characterized by frequent floods, open sewers, industrial waste, trash, dead fish, old tires, and oil slicks, a health hazard for anyone swimming in it. There was little visual connection between the river and the city—in fact, the river bisected the city.

The report suggested several beautification projects as well as new development. To make the area more beautiful, the report recommended screening the industrial areas on the east side of the river with landscaping, undertaking a major cleanup of trash and debris, and protecting the riverbank from further erosion. For new development, the report suggested a major waterfront park, a park on the east side of the river to honor Toledo's ethic heritage, residential areas along the river, recreational facilities (including public marinas), and a pedestrian walkway along the riverfront. In recognition of Betty Mauk's work, the plan said any development must incorporate Promenade Park. The plan promoted retaining several older structures with historical significance, including the Toledo Edison Steam Plant, the Fort Industry Block, the Bostwick-Braun building, the Waldorf Hotel, the Securities and Exchange Building, and the Monroe Street bridge over Swan Creek. New buildings were also a part of the plan, including a city/state office complex, a corporate office tower, a new hotel, residential apartments, a community conference center and sports complex, and possibly a new museum honoring the city's history as a center of the glass industry.

Given the previous failure with the convention center, the GTC endorsed the plan, but pointed out that its success would depend on developing a sound and lasting consensus between public and private leadership.[14] This required the involvement of the right people, the support of appropriate organizations, and realistic plans for public and private investment. The GTC hoped that by working with the TDC, such consensus would be reached, and the money would be found to expedite the vision laid out in "Toledo Looks to the River."

Despite these warnings and two years of planning, the first step of the GTC to realize the plan stumbled badly. Seizing on the suggestion for a community conference center along the river, the GTC allocated $60,000 for a plan to build a multiuse building for community activities, including a new building for the University of Toledo's adult and continuing education program. The plan also included a significant presence by Owens-Corning Fiberglas, which agreed to lease a substantial portion of the new building for some of its corporate functions.

But the plan soon ran into a major obstacle—Toledo *Blade* publisher Paul Block Jr. In June 1975, Mayor Kessler had appointed Block to the position of chair of the TDC in recognition of Block's long interest in downtown development that dated back to the "Toledo Tomorrow" exhibit. Block's appointment was not met with unanimous support, as many citizens felt that being in such a public role conflicted with Block's ownership of the *Blade* and his control of its editorial pages.[15] But Edwin Dodd supported Block's appointment, and this signaled an end to a contentious battle that had been waged on the *Blade*'s editorial pages with Owens-Illinois over the naming of the library at the Medical College of Ohio in honor of Raymon Mulford, the recently deceased president of O-I. Block, who had been the force behind the establishment of the medical college, had been angered by the choice of Mulford as the library's namesake. With Block's appointment as chair of the TDC, the battle with O-I temporarily ceased.

Almost immediately, problems with the multiuse building began. One month after being appointed chair of the TDC, Block proceeded to develop his own downtown redevelopment master plan. To design the plan, he hired Minoru Yamasaki, an internationally known architect with offices in Troy, Michigan, who had designed the World Trade Center in New York and whom Block had worked extensively with in designing buildings at the Medical College of Ohio.

Unlike the "Toledo Looks to the River" report, Yamasaki's plan did not include a multiuse community building on the riverfront. Some members of the GTC were concerned that Yamasaki's plan was directly contradictory of previous plans, disregarded existing viable buildings, and did not consider downtown traffic flow patterns. The GTC leadership, recognizing the collision course it was on with the TDC and its powerful leader, met with Yamasaki in September 1975 in an attempt to bring him into the project.[16] The offer was declined.

Because of Block's opposition and his ownership of the most powerful local media outlet, the UT board of trustees began to get cold feet about the adult and continuing education building. UT's leaders had already tangled publicly with Block on the medical college, which was established as a standalone institution rather than as a part of the University of Toledo, as originally planned. Despite the apprehension of some, the UT board tentatively agreed on December 18, 1975, to participate in the multiuse building project.

Other business leaders lobbied hard for the TDC to support the project. In a letter dated December 4, 1975, Leslie J. Barr, president of the GTC, wrote to Mayor Kessler and Block to clarify the project's components.[17] On the top floor would be offices overlooking the river, there would be many multipurpose community meeting rooms, and the Adult and Continuing Education Center would actually be a separate structure connected to the building. The multiuse building would also include restaurants and specialty shops. William Boeschenstein of Owens-Corning Fiberglas wrote to the mayor on December 30 on the eve of the TDC's vote and stated his company's commitment to lease 75,000 to 100,000 square feet of the building.[18]

Despite these efforts, on December 31, 1975, the TDC deadlocked over giving its approval to the project by a six to six vote, with Block leading the opposition. Two days later, the Citizens' Development Forum, the citizens' advisory panel to the TDC, voted to support the project. But the project went to the city council without the backing of the TDC.

Opposition also mounted on the UT campus. With $2.5 million of the money to build the building coming from the state of Ohio's capital improvement fund and another $800,000 from the university's Centennial Fund Drive (a major capital campaign that marked UT's 100th anniversary), UT's participation was required if the project was to proceed. Major supporters of the Centennial Fund Drive, particularly R. A. Stranahan, said he opposed using money raised under the pretense of building a continuing education center on UT's campus for one that was to be built downtown.[19] In early 1976, a month after the TDC voted against the project, the UT board of trustees withdrew its support. President Glen Driscoll said the university risked losing the state capital improvements money for the Adult and Continuing Education Center because of the delays created by the debates over the project.

After the failure of the multipurpose building, the GTC created another committee called the Toledo Economic Planning Council. The job of the council was to find federal financing that might be available for downtown redevelopment since local and state funding could not be counted on. George Haigh, president and chief executive officer of Trustcorp, Inc. (the holding company of the Toledo Trust bank), was appointed chair of the council.

The "S.O.B."

Following the disappointment of the multiuse building, Edwin Dodd of Owens-Illinois declared the GTC would continue to push forward with downtown redevelopment. The GTC had learned a valuable lesson, however: do not move forward with any projects without the support of the TDC and Block. The next proposed redevelopment project went forward with the strong support of both groups. Yet it too did not proceed smoothly. This time, opposition came from an unlikely source—the congregation of one of Toledo's oldest churches.

With Yamasaki's interest in the downtown, the GTC and the TDC began discussing the feasibility of and location for a new state office building as a "keystone of public investment in the emerging downtown."[20] In

June 1977, Block, Dodd, and Leslie Barr, along with other members of the GTC and TDC, met in private in Troy at Yamasaki's headquarters to discuss his proposed plan for the downtown.[21] The plan was revealed by Block one month later. It showcased a new state office building across Erie Street from the county courthouse, placement Yamasaki stated was necessary so as to create a civic plaza. However, building the structure there would require moving the 110-year-old St. Paul's Lutheran Church one block north, the cost of which would be included in the construction costs for the new state office building, which were then projected at $25 million. Yamasaki's plan also called for demolishing several other buildings near St. Paul's that housed retail shops and law offices. Also to be demolished were the Lion Store, the Renaissance Municipal Building, the former headquarters of Toledo Edison, the Civic Center garage building (which was also occupied by WTOL-TV and two radio stations), the Gardner Building, the Lamb Building, and the Waldorf Hotel. Despite these obstacles, on July 28, 1977, the TDC unanimously approved Yamasaki's plan. Libbey-Owens-Ford jumped on the Yamasaki plan, proposing in August a $1 million face-lift for the area around its headquarters that reflected Yamasaki's ideas.

But plans to move St. Paul's met with considerable grassroots opposition. In a letter to Mayor Kessler, Vernon Rohrbacher of the law firm Cobourn, Smith, Rohrbacher, and Gibson, whose offices would be demolished for the state office building, registered his complaints in the "strongest possible terms."[22] Rohrbacher called the plan for the building ill-conceived, and pointed out that the plans did not include an estimate of the cost of moving the church and that they were not based on a single feasibility study. "The inclusions into the plan of such ideas as razing numerous well-maintained and aesthetically pleasing structures and the moving of the landmark St. Paul's Lutheran Church marks the plan as one which has as its main function the further promulgation of the dictatorial voice of Paul Block, Jr. Even though strong attempts at obfuscation have been made, Mr. Yamasaki has admitted on several occasions that the site for the city-state building was chosen for him, rather than by him."[23]

Ballots were sent to the members of St. Paul's in January 1978 to vote on whether the church should be moved. Block's newspaper urged support for the move, stating that other downtown redevelopment efforts—including a new world headquarters building for Owens-Illinois—depended on approval by the congregation. Other news outlets, including WSPD-TV 13, expressed strong opposition. The editorial stance of the television station struck a chord with some, including powerful business leader Stephen

Stranahan. In a letter to David Drury, WSPD's editorial director, Stranahan said, "To ask a 2,200-member church to be a pawn in this architectural extravaganza hatched by Block and Yamasaki is very inconsiderate of the role the church has played in its 120-year history in this location, where it has tried to stabilize an area but complement its governmental surroundings. To heap on top of this the rudeness the inferred success or failure of mostly unrelated and somewhat distant building development projects is wantonly unfair and smacks of bossism here in Toledo that even New York City eliminated some years ago with the retirement of Robert Moses."[24]

St. Paul's congregation clearly expressed itself on the matter when its votes were tallied on January 22, 1978. The members opposed moving the church with 293 in favor, 784 opposed.

Mayor Douglas DeGood, who had been elected to replace outgoing mayor Kessler the previous November, continued Kessler's backing of the plan. On February 21, 1978, DeGood reiterated Block's contention that St. Paul's relocation was the only stumbling block to redeveloping the entire downtown, including a new headquarters for Owens-Illinois. DeGood hinted he would approach the church with other offers that would allow the Yamasaki plan to move forward. The congregation was asked to consider another option—building a new St. Paul's near its current location but one out of the way of the state office building (now often cynically referred to by some in the city as the S.O.B.).

But the congregation would have none of it. Having saved the church from being moved, the congregation was in no mood to see it demolished, even if the estimated $3 million to build the new church would come from the state. On April 30, 1978, the members turned down the offer for the new church. Paul Block called the decision unfortunate. Edwin Dodd said the decision by the church members might delay his grand plans for Owens-Illinois's new headquarters.

By May, the TDC had resigned itself to a new location for the building, one bounded by Jackson, Erie, Beech, and Huron streets. The group voted unanimously in favor of the new location after Edwin Dodd expressed his support for it. In September, the Ohio Building Authority voted two to one to acquire this land, even though costs for the building had escalated from $25 million to $60 million. Yamasaki was hired as the architect.

Once the state approved building the county-city-state office building, Paul Block announced on May 25, 1979, that he was resigning as chair of the TDC. Mayor DeGood dissolved the TDC five days later, stating "there was a general feeling that the committee had successfully gotten downtown revitalization underway. And the best thing to do now was to dissolve it."[25]

The 22-story city-county-state office building was completed in 1983. It was but one part of a grand effort to remake the downtown.

The Dawn of SeaGate

Owens-Illinois had moved into its headquarters on Madison and St. Clair in 1935, just six years after the company had been formed. By 1969, the company began to question how long the building would remain viable for its corporate offices. That year, the company conducted a long-term study of space needs and concluded the Madison building would need to be replaced by 1981, and urged a move to the suburbs. Two years later, the "Toledo Area Facilities Study" concluded the company would need much more space by 2000, but rather than leave the downtown, O-I could build a second building there, expand office space at its Levis Development Park in Perrysburg, or expand offices at the Technical Center on Westwood Avenue in west Toledo.[26] In 1973, the company began to buy up property around its headquarters building to give it space to expand, acquiring 9 of 11 buildings on the block. Yet another study in 1976 suggested a brand-new building of nearly 800,000 square feet, large enough to house all 2,600 administrative employees predicted for the company by 2000.[27] One location considered for a new building was the site along the riverfront of the former Tiedtke's Department Store, a Toledo landmark that had burned to the ground in a sensational fire in 1975.

In May 1977, company executives met with Mayor Kessler to articulate what was required to keep Owens-Illinois—the city's largest corporation—downtown. The mayor, recently stung by the failure of the multiuse building, announced at a meeting of the TDC on July 18 that a successful downtown redevelopment would depend on more than public investment. "It is now time to abandon our passive attitude about stimulating private investment in downtown Toledo. Long and frustrating delays have created a climate of uncertainty for the potential downtown developer, as well as increasing costs. . . . No one from Houston, New York, or Chicago is going to invest in downtown Toledo unless we are first able to convince northwest Ohio corporations and investors that city government is committed to assisting privately sponsored downtown revitalization programs. I want to emphasize this as emphatically as possible. If Toledo is to develop a positive momentum in attracting other national corporations to this city we must begin with the basic strength of our own existing corporations."[28]

At that TDC meeting, the group revealed its plans to offer O-I the 11-

acre tract of land along the river where Tiedtke's once stood. The land was held by the city, and had been acquired through a series of urban renewal grants from the federal Housing and Urban Development office. If Owens-Illinois agreed to purchase the land, the sale would have to be approved by HUD. Another contingency was that if O-I did agree to purchase the land for a new headquarters building, O-I would require Toledo to make extensive "public" improvements to the site.

Another big hurdle for the deal was that the city would also have to agree to a 20-year property tax abatement. If the city agreed not to collect property taxes, Owens-Illinois argued the deficit would be offset by city income taxes paid by those working in the new building and by rising property values resulting from the development.[29] On September 29, 1977, Toledo City Council approved the sale of 10.8 acres of land for $2.1 million to the Maumee Valley Urban Redevelopment Corporation, a subsidiary of O-I.

While Edwin Dodd was ready to go forward with the new headquarters, the O-I board of directors still needed to be convinced. In a speech in October to the board, Dodd focused not on the altruistic desire to help the city, but rather on the economic reasons the move would be good for the company. "Beyond such considerations of corporate citizenship, some very real benefits would accrue to Owens-Illinois from the revitalization of the downtown," Dodd argued. "It may be tempting to ignore the deterioration of the downtown as we concentrate on the narrowly defined demands of our business. But it is obvious to anyone willing to face reality that the vitality of the community where our world headquarters is located is an important factor in our business."[30]

O-I's influence on its corporate neighbors was quickly felt, with Toledo Trust, one of the city's largest and oldest banks, announcing in January 1978 that it too would build a new headquarters downtown on the site of the old Waldorf Hotel if O-I moved forward with its plans. The Waldorf Hotel had been one of the historic buildings that the "Toledo Looks to the River" report urged the city to preserve.

While everything seemed in place for the project, previous redevelopment efforts had shown that the smallest obstacles could prove fatal. For the project to move ahead, Toledo had to acquire federal funds for necessary public improvements such as lighting, a parking lot (which was to include a heliport Dodd contended was needed for corporate helicopters), and street beautification. The city requested an $18 million Urban Redevelopment Action Grant, but received only $12 million—still the largest UDAG grant awarded that year. City fathers noted that the success of the grant application was due to the cooperation of federal, state, and local governments with private industry and community groups. Unlike previous redevelopment efforts, O-I's project was moving smoothly.

Throughout 1978, plans for the building began to take shape. That year, Edwin Dodd was named "Citizen of the Year" by the Toledo Board of Realtors because of his commitment to the downtown. But some members of O-I's board of directors believed the initial cost estimates of $93 million for the new headquarters were too high, so cuts were made to bring the cost down to $88.5 million. At the same time plans for the building were proceeding, rumors began circulating among headquarter employees that some would have to be laid off to pay for the building.[31] Another group expressing concern about the building was the Toledo Board of Education, which estimated the tax abatement would cost the system $1 million a year for the 20 years of the agreement. Yet another opponent of the tax abatement was Carleton Finkbeiner, a rising political star in the city. Despite the opposition, the tax abatement plan was unanimously endorsed by city council after a deal was struck that allowed O-I to receive the tax abatement while Toledo Trust withdrew its request for a similar deal.

The *Blade*, in an editorial in January, called 1979 "The Year of Promise." "As Toledo heads into another new year, it can do so with a good deal of confidence that the pendulum of progress is surely swinging to the positive side. That is especially true as it relates to the major effort that is approaching to redevelop a significant section of the downtown area," the *Blade* pronounced.[32] The name SeaGate was selected as the overarching name for the redevelopment projects that encompassed the Owens-Illinois project.

The new Owens-Illinois headquarter was to be a 32-story glass skyscraper designed by architect Max Abrahamovitz. In a talk to the American Institute of Architects in May 1979, Abrahamovitz stated, "We think when this is done properly and finished its going to be one of the most interesting environments for an office building and a complex on the river that has ever been done in this country. . . . There are other cities that have some water development but I don't think they are going to compete with this."[33]

In order to build the O-I headquarters building, the bank, and other projects that were a part of SeaGate, several historic structures had to be demolished. These included the Waldorf Hotel and the Securities and Exchange Building. An attempt was made to preserve the facade of the latter, which had been determined by the National Advisory Council on Historic Preservation to be a building of historic significance. But as demolition began, the weakness of the building became obvious, and the marble facade

The groundbreaking ceremony for the new headquarters building for Owens-Illinois, 1979. *From left to right:* Edwin Dodd, Thomas Ludlow Ashley, Mayor Douglas DeGood, and Paul Block Jr. (Ward M. Canaday Center for Special Collections. Used by permission of Owens-Illinois.)

"crumbled like a stale wedding cake."[34] In 1981, the Board of Trade building, an imposing structure at St. Clair Street and Jefferson Avenue that had stood since 1925, was demolished to build a parking garage adjacent to the Owens-Illinois building.

As plans were finalized, O-I symbolically stated its move from its old building by excising the corporate logo that had stood atop it and had marked the Toledo skyline for decades. Parts of the sign—including the large "Owens" and "O"—were removed by helicopter and donated to Owens Technical College.

Groundbreaking for the new headquarters took place three days later on May 22, 1979. Called "The Start of Something Big," the event was meant to showcase the community's involvement. The *Blade* reported the event was marked by "a contagious enthusiasm for Toledo's future."[35] O-I invited thousands to participate by turning a shovel of dirt after CEO Edwin Dodd,

Removing the O-I sign from the company's first headquarters building, 1979. (Ward M. Canaday Center for Special Collections. Used by permission of Owens-Illinois.)

Congressman Thomas Ludlow Ashley, Mayor DeGood, and Paul Block Jr. turned the first shovels. Nearly 6,000 people, many of them O-I employees, participated. O-I hired an accounting firm to verify the number, and submitted the event for consideration to the Guinness Book of World Records, hoping to be recognized as the largest groundbreaking ceremony. Unfortunately, the Guinness company refused to consider the event, stating in a letter to O-I executives, "As you yourself have already discovered, competition data on such an occurrence is scarce and this is one of the prerequisites for acceptance of record claims, we do hope you will understand why we are unlikely to be able to include an item of this kind in the Guinness Book of World Records."[36]

One SeaGate, the headquarters for Owens-Illinois from 1982 to 2006. (Ward M. Canaday Center for Special Collections. Used by permission of Owens-Illinois.)

The new building, when completed in 1982, was the tallest building (taller even than the Fiberglas Tower) in Toledo, a 411-foot glass and steel skyscraper. The project included a restaurant operated by noted Toledo restaurateur Dick Skaff, a retail complex, an underground connector to the parking garage across the street, and eventually a glass-enclosed walkway connecting several downtown buildings on Summit Street.

Make No Small Plans

The state office building and the new O-I headquarters were just the beginning of the grand plans for transforming Toledo into a city that would no longer be the subject of satirical songs like "Saturday Night in Toledo, Ohio." As early as 1975, Leslie Barr, president of the GTC, began looking at the Faneuil Hall–Quincy Market area of Boston as a possible model for retail development in Toledo.[37] In Boston, developer James Rouse of the Rouse Corporation spent over $20 million to lease and renovate 44 derelict buildings in the center of the city to create a unique retail center focused on small specialty shops, bars, and restaurants. Barr felt Toledo could duplicate the development, perhaps using the Old Town area near the riverfront where the few original historic downtown buildings that still remained standing were located.

In June 1978, the city sponsored a forum called "What is a City?" held at the Trinity Episcopal Church. Congressman Thomas L. Ashley urged the city to spread revitalization to other areas of the downtown beyond the SeaGate area. Also speaking at the forum was James Rouse, who told the audience that Toledo did not yet have the infrastructure to support a "festival marketplace" like Quincy Market, but that it did have enormous potential.[38]

But it was also clear that Toledo had enormous problems. Just two months after the spectacular groundbreaking for SeaGate, the Holiday Inn, the only major downtown hotel, was taken over by five local banks due to financial problems. In September 1979, the YMCA confirmed it was trying to sell its downtown building, one of the architectural gems of the city, and it would move out of the downtown by October. A group quickly came together to try to save the building, and the YMCA was convinced to keep the building open through June 1980. The Lamson's Department Store, which had declared bankruptcy and closed in 1976, found new life as an office building. In 1979, Owens-Corning Fiberglas moved some of its offices into that building, renaming it One Lake Erie Center. But the Lion Store was not so fortunate. On November 28, 1979, the store, one of the original Toledo department stores with roots dating back to 1857, announced it was closing its downtown building in February. The building would eventually be torn down. LaSalle's seemed destined to follow. On December 11, city manager Michael Porter said city finances were in trouble due to the downturn in the automobile industry, and the budget of the city that had been built upon estimates of city income tax receipts would have to be revised.[39]

The "Year of Promise" that the *Blade* had declared was ending with less than promising news.

Despite these realities, redevelopment marched on, with the second phase of the SeaGate development announced in early 1980. It would include a new $30 million hotel to be built adjacent to the O-I headquarters and developed by John A. Galbreath and Company of Columbus. Like the Owens-Illinois headquarters, the hotel would also require a tax abatement to go forward. Proponents noted that the hotel would increase the ability of Toledo to attract conventions, would employ 350 people, and contribute $4 million to the local economy in exchange for the $400,000 abatement.[40]

But plans for the hotel were met with opposition from a familiar voice, Paul Block Jr. Block noted that the hotel was not in keeping with the Yamasaki master plan, and would cut off the river from the public and limit open spaces in the downtown. Yamasaki himself weighed in on the decision, stating in a letter to the editor of the *Blade*, "To throw away the last opportunity to make Toledo a lovelier place in which to work and live seems to be a great shame. As an architect, I advocate that the people of Toledo do their best to eliminate this proposed building and instead build a fine waterfront park between the O-I Building and the Toledo Trust headquarters."[41] As the *Blade* summed up the argument, "The land, it is true, belongs to Owens-Illinois—but the river belongs to the people of Toledo."[42] O-I countered that the hotel would be designed in a way that provided for more open land than even the "Toledo Looks to the River" plan of 1975 envisioned. City Council took up the matter after Betty Mauk, the community activist who had worked for years to protect the riverfront, led a petition drive to stop the project.

At the same time Owens-Illinois leadership was promoting the hotel, the company closed its bottle plant in Perrysburg, and many employees believed the action was due to the cost of the downtown projects. O-I countered that it was the result of a mandatory bottle deposit law recently passed in Michigan that reduced the need for new bottle production.[43]

While the *Blade* and O-I had cooperated on the SeaGate project, that alliance came to an end in the summer of 1980 as O-I moved forward with its hotel plans. In an editorial entitled "A Time to Say No," the newspaper expressed appreciation for O-I's commitment to the downtown.[44] But it also reminded readers that the company had received a 100 percent tax abatement for 20 years for its headquarters, had demanded and received 10 acres of prime riverfront property for only $2.1 million, and had forced the city to make major improvements to the area, including the construction of public marinas, a parking garage, a pedestrian walkway, and street improvements.

Now the company was requesting that the city seek an additional $7.5 million in federal grants to help build a hotel next to the new headquarters building. "The time has come, it seems to us, for Council to politely but firmly say no," the *Blade* stated. "Placing a hotel on that particular site—forever shutting off the riverfront from much of the downtown—would be a classic case of poor planning."[45] Council unanimously approved the federal grant request on July 28 despite the *Blade's* opposition. Plans for the hotel moved forward, and the Hotel Sofitel, part of the upscale French chain, opened in 1984.

Despite losing the battle with Owens-Illinois over the hotel, in November 1981 the *Blade* and its publisher Block touted the success of downtown redevelopment as the realization of the dream that had been expressed in 1945 in the "Toledo Tomorrow" exhibit. "The 1945 'Toledo Tomorrow' exhibit did help put Toledo firmly on the path of civic progress. It is not important that the 1945 model was not ultimately followed. What was important then and remains important now is that the city's residents maintain a vision of their community that is always somewhat beyond their grasp. The urban progress story does not really ever end; there must always be a challenge to be grappled with tomorrow and the day after and the day after that."[46]

And the dream continued to march on. The second phase of SeaGate also included a festival marketplace—also located near the O-I headquarters—designed by James Rouse's company, Enterprise Development. It was one of several such developments that Rouse undertook in smaller cities, including Flint, Michigan, and Norfolk, Virginia. Opened to great fanfare in May 1984, the mostly glass pavilion was designed to create a carnival atmosphere with food kiosks, restaurants and bars, and shops so specialized that some were devoted exclusively to kites, fancy hats, items depicting pigs, and purple merchandise. Much was riding on the success of Portside. As *Blade* columnist William Brower stated in his column on the day of its opening, "So much has been staked on the new venture that failure is considered out of the question. Anything that would breathe new life into the business core—commercially, esthetically or gregariously—should be a welcome phenomenon. . . . If Portside cannot lead the way to a reawakening of downtown merchandising and entertainment, Toledo is in deep trouble. This isn't a pessimist speaking; it is the voice of hope saying this is it," Brower stated prophetically.[47]

Developers expected Portside to draw five million visitors a year, generate $18 million in sales, and create 700 jobs.[48] Perhaps an omen of things to come, however, Macy's (formerly LaSalle's), the last department store in downtown, closed two months after Portside opened. Visitors to the downtown could no longer purchase clothing or shoes from a department store, but they could buy purple socks or items shaped like a pig.

The massive new development of the downtown was underwritten through a financing scheme described as "the most Byzantine financing I have ever seen" by one anonymous Toledo official quoted in the *Blade*.[49] Some observers felt it was a house of cards, with one new project serving as collateral for the next, and Toledo Trust holding most of the mortgages. Any change in the economic underpinnings of the city had the potential to be catastrophic for the entire downtown.

But there was no stopping the development snowball. To draw more visitors to support the attractions, in 1987 a new convention center was finally realized with the opening of the SeaGate Centre, a joint project of Lucas County and the University of Toledo. Using a $10 million allocation of state capital improvement money, the convention center also included classrooms so downtown workers could further their education at UT. When the SeaGate Centre opened in 1987, it already had a $1.3 million operating deficit.[50]

Like Being Nowhere At All

The precariousness of the downtown development plans began to become clear in late 1986 when the giant holding company Kohlberg Kravis Roberts & Company approached Owens-Illinois CEO Robert Lanigan (who had replaced Ed Dodd) and presented a buyout plan to take over the corporation. While the company initially fought off the takeover bid, by February 1987 it succumbed, at a cost of $3.66 billion. As was typical of such takeovers, the cost of the buyout was financed by the assumed company's assets, which were quickly sold off to pay the massive debt. With its reduction in size, the number of Toledo employees of Owens-Illinois dropped from 5,600 in 1980 to 3,400 in 1990.[51]

Corporate restructuring and the recession of the late 1980s reduced Toledo's workforce significantly. Few Toledoans had the disposable income to support the downtown attractions such as Portside, and it began to fade almost as quickly as it was built. By 1989—just five years after it opened—over half of the businesses at Portside were empty. It closed entirely the following year.

The changing economy and the rush of corporate buyouts also impacted the local banking institutions. First Federal Savings and Loan fell victim to the mismanagement that characterized the savings and loan scandals of the time, and was taken over by the Resolution Trust Corporation and eventually sold to a Cleveland bank.[52] Toledo Trust was forced to write off $75 million in bad loans, most of them for downtown redevelopment projects.[53] George Haigh, who had led much of the bank's efforts at downtown investment and who was a major city powerbroker, was forced to resign his position, and the bank was sold to Society Bank of Cleveland in 1990.

Within one brief decade, Toledo's downtown had seen its great hopes for the future quickly dashed, and the domination of its glass industry began to fade. In an editorial appearing in the *Blade* on January 1, 1990, the newspaper, which had championed a revitalized downtown since 1945, soberly faced reality. "At this year-end juncture a decade ago, Toledo was a city flush with an exhilarating sense of rebirth. Its old downtown was about to undergo not just a facelift but major reconstructive surgery. . . . The giants were in the driver's seat, [and] the corporate leaders whose combined financial muscle and will to move Toledo forward had joined to promote an ambitious and grand plan for the heart of the city. Today, the giants, for all purposes, are out of it; corporate financial might has been depleted or rerouted; the ambitious scheme—Fort Industry Square, Portside, the Sofitel-now Marriott, the Radisson, SeaGate Centre, Summit Center, a new retail corridor on Adams Street—was laid flat by a leveraged buyout uppercut," the editorial lamented.[54]

In many ways, Toledo has yet to recover from the psychological blow resulting from the failure of these grand downtown redevelopment schemes. The failure of downtown redevelopment also signaled the coming end of Toledo's domination by the glass corporations.

NINE

Cracks in the Glass City

> The remarkable growth of our city during the past century has been one of the marvels of the world. Starting 111 years ago as a little country village of 100,000 inhabitants it has steadily grown to its present population of over 6,000,000. Only one city, New York, now rivals it in numbers. London, Paris and the cities of the east have one after another been left behind.
>
> —Prediction of Toledo in 1999, from the *Toledo Daily Blade*, January 12, 1888

In May 2003, Toledo was faced with the prospect that all three of its largest glass companies might exit the downtown at the same time. The circumstances that led to this devastating possibility are complex, and in many ways epitomize the forces that have shaped the recent economic history of the United States. A vortex of globalization, corporate greed, mismanagement, costly litigation, bankruptcy, workforce downsizing, and changing views of corporate citizenship swept the city and threatened its very foundation.

That month, Owens-Illinois, Owens Corning, and Pilkington PLC (successor to Libbey-Owens-Ford) each threatened to leave downtown Toledo. Owens-Illinois, which had invested nearly $100 million in 1982 to build its gleaming One SeaGate building that was supposed to stimulate the revitalization of the entire downtown, decided under new corporate leadership that the skyscraper no longer fit its culture. It also no longer fit the size of the company's downtown workforce. While One SeaGate was built to house 2,200 O-I employees, in 2003 only 340 remained.[1] Pilkington PLC, the British company that had taken over Libbey-Owens-Ford in 1986, found its headquarters on Madison Avenue was also too large for its downsized corporate staff. The company no longer needed to occupy 12 floors—4 were sufficient.[2] It began to look for another location more suitable for its size. Owens Corning, which had vacated the Fiberglas Tower in 1996 for a 400,000-square-foot building on a campus along the Maumee River, declared bankruptcy in 2000 under the weight of costly asbestos litigation. It claimed it could no longer afford the rent on its new building.[3] The Fiberglas Tower remained vacant seven years after the company had moved out, adding another eyesore to the downtown.

If all three companies had moved out of the downtown at the same time, it would have left Toledo's inner core a ghost town with thousands of feet of vacant—and worthless—office space.

In the end, only Owens-Illinois moved on to greener, suburban pastures. But to understand how the city came to the brink of disaster, one must understand the events of the past 30 years and how they revealed the glass industry in Toledo to be as fragile as the product it produces.

The British Invasion

The years 1986 and 1987 were ones of shake-ups in the Toledo glass industry. Owens-Illinois was not the only Toledo glass company to experience dramatic change. In March 1986, Libbey-Owens-Ford was sold for $353 million to the Pilkington Brothers, PLC of Great Britain.[4] The Pilkington company had been a pioneer in plate glass making, with roots dating back to 1826, when it had been founded as the St. Helens Crown Glass Company in St. Helens, England. In 1891, Edward Ford had toured a Pilkington factory to gain more knowledge about plate glass production that he then took back to his father's company, Pittsburgh Plate Glass. In his diary, he noted how impressed he was with the facility and the production methods used there.[5]

In 1983, Pilkington first expressed its interest in L-O-F by purchasing over 3 million shares of the company's stock. Three years—and 95 years after Edward Ford had toured its factory—Pilkington bought the rest of the company that Ford had created in Rossford in 1899.

The 1986 sale included not only the Rossford plant but also L-O-F's glass research facility on East Broadway. That facility started as a laminated glass factory, built by Edward Drummond Libbey and Michael Owens in 1921.

The sale also included the corporate headquarters building built in 1960 on Madison Avenue and the L-O-F Modern Tools Division, which was located in north Toledo. Pilkington chose to keep the Libbey-Owens-Ford name because of the value of the brand name.[6]

Once Pilkington took over the company, it immediately split off the nonglass divisions that L-O-F had acquired in the 1970s into a new company. These divisions included Aeroquip, which made flexible hoses and couplings; Vickers, which produced fluid hydraulic systems; and Sterling Engineered Products, which made plastics. After much consideration, the name Trinova was selected for this new company, for the three pillars of its products.[7] Trinova, which was headed by president Darryl Allen after the retirement of Don McKone, moved its headquarters to a new campus in Monclova, about 15 miles west of downtown Toledo. It quickly went on a buying binge of its own, purchasing engineered product companies, aerospace-related manufacturers, and more plastics plants. The company had no interest in the glass business anymore. The split between Trinova and Libbey-Owens-Ford was mostly amicable except for a lawsuit over pension payments.

Pilkington expressed its commitment to keeping Libbey-Owens-Ford in Toledo and its production facilities in Rossford, which was a relief since many feared it might move the headquarters to England. This was desirable for the company because of the close proximity to the company's major customer, General Motors, in Detroit. In all, Pilkington took over 13 plants and 7,400 L-O-F employees.[8]

In 1993, Libbey-Owens-Ford was back on the front pages of the local media when news broke of a scandal involving its top executives. Pilkington removed president Ronald Skeddle along with vice presidents Darryl Costin and Edward Bryant over charges that the three had defrauded the company of $14 million.[9] The scandal resulted from a corporate directive. In an effort to cut costs and increase profits, Pilkington had urged leaders to outsource some activities, but claimed to be unaware that these executives had personal stakes in some of the firms performing the contracted work. The initial 274 counts of defrauding the company against the three were reduced to 20, and after a three-month trial, all three of the former executives were acquitted.[10] A civil case was settled out of court. Reflecting on the events 10 years later, Pilkington spokesperson Roberta Steedman said that the company emerged from the scandal stronger, and that its focus on corporate governance issues had put the company ahead of others.[11]

Hostile Takeover Attempt

In addition to Libbey-Owens-Ford and Owens-Illinois, Owens-Corning Fiberglas was also the subject of a takeover attempt. In August 1986, Sanford L. Sigoloff, president of Wickes Cos., which specialized in lumber and home improvement products, informed Owens-Corning that Wickes was attempting to buy it. For several months, Wickes had been quietly buying up Owens-Corning stock and, by the time it attempted its takeover, already owned 8–10 percent, with an option to purchase 2.2 million more shares.[12] The management of Owens-Corning had observed the stock trades, but did not know who was doing the purchasing. The trading in Owens-Corning stock had driven the price up, from about $50 a share to over $60. The takeover offer Wickes made was for $70 a share.[13]

William Boeschenstein had feared such a takeover attempt, and company managers had begun to develop a strategic plan just in case. When it happened, Boeschenstein did not take the Wickes move well, and some believed he took it personally because of his father's role in creating the company.[14] In a press release issued the day after the offer by Wickes was made, Boeschenstein expressed his contempt carefully but strongly, and countered with a business restructuring plan that would produce short-term gains for stockholders. "We strongly believe that the current management team, which knows our business intimately, is far more qualified than any outsider to assess and implement any such [business restructuring] program," Boeschenstein said.[15] He also charged that Wickes may have violated the Hart-Scott-Rodino Antitrust Improvements Act of 1976, which required companies to publicly disclose their intentions to take over a company before making any large stock purchases.

Amazingly, Wickes had only the year before emerged from bankruptcy. But under the aggressive leadership of Sigoloff, the company went after other companies with good cash flows. It was particularly interested in Owens-Corning's insulation and roofing products, which fit well with its lumber business. Within a few days, Wickes upped its offer to $74 a share.[16] That offer was rejected by Owens-Corning as well. Boeschenstein again stated that he wanted to restructure the company rather than accept a buyout. In an oral history interview recorded in 2001, Boeschenstein recalled, "At no time would we have considered selling the company to Wickes. It would have been a disaster for the business, and for the people because they had no idea, we didn't think, of how to run the business."[17]

But while Boeschenstein was publicly protecting the company, a blind trust of the Boeschenstein family as well as several individual Owens-Corning executives sold off large blocks of company stock when its value reached $76 a share due to the Wickes offer.[18]

By August 30, less than a month after its initial offer, Wickes announced it was ending its effort to take over Owens-Corning Fiberglas. Because the stock price had soared during that month, Wickes still walked away with $9.4 million in profits.[19]

Owens-Corning, on the other hand, saw no profits from the takeover attempt, and instead was mired deeply in debt. To defeat Wickes, Owens-Corning's executives offered stockholders $52 in cash for each share of stock owned as well as a $35 20-year bond. The total cost of this offer? Some $1.8 billion.[20] To fund the offer, Boeschenstein had to dramatically resize and restructure the company. Almost immediately, rumors of massive layoffs began to circulate among employees at the headquarters in Toledo as well as at it major manufacturing and research facilities in Newark and Granville, Ohio.

To announce it had survived the takeover attempt, Owens-Corning published a full-page advertisement in the *Blade* on October 5 that declared, "It's Full Steam Ahead at the New Owens-Corning." The ad stated that the company would narrow its focus to its core businesses which were most profitable, and would be "running a lean, tight organization."[21] The advertisement went on to note that this would come at a considerable cost. "Certainly, there will be some pain—especially in the reductions we must make in our work force. But we intend to ease this process wherever possible. . . . The toughest part is still ahead of us—to make our plan work. And here again, we will need the combined efforts of thousands of Owens-Corning people, and the continued support of all of you whose lives we touch," the advertisement concluded.[22]

The business restructuring plan reduced eight divisions to three. The downsizing included the company's Aerospace and Strategic Materials Group, which it had just purchased the year before from Armco, Inc. The company also sold off its ceiling products operations, its foam sheath and roof insulation manufacturing plants, and its reinforced plastics business that made bathroom shower components. It also dramatically reduced its research-and-development efforts. The cutbacks hurt the community of Newark, Ohio, particularly hard, with layoffs of between 700 and 900 employees. Other major reductions occurred in Kansas City, Kansas, were 350 were laid off; Santa Clara, California, where 250 lost their jobs; and Barrington, New Jersey, where 650 employees were let go.[23]

In Toledo, Boeschenstein announced in October 1986 that 650 jobs would be cut from the headquarters staff of 2,000.[24] The reaction was, as might be expected, one of paranoia and declining morale. Some Owens-Corning employees took their vengeance out on Sanford Sigoloff of Wickes, writing letters expressing their disgust with the takeover bid that seemed only to benefit the executives of Wickes. "For whatever reasons, you attempted an 'unfriendly' take-over of OCF. Whether it was a ploy to make money for Wickes, or if you really intended to take us over, I think you should be aware of the devastation of so many people here. . . . I understand the business world, Mr. Sigoloff. I understand what it takes to keep a company—yours as well as mine—on its feet. But there is a fine line between right and wrong. I believe that word is 'ethics,'" one employee angrily stated.[25]

The outcome of the restructuring was startling. Sales were projected to drop by $1 billion with the reduction in product lines.[26] The total number of employees was projected to be cut from 28,000 worldwide to 15,000. And the company's debt soared from $567 million to $2.62 billion. The cost to Wickes? Just a $300,000 fine levied by the Justice Department for violations of the Hart-Scott antitrust law.

While Owens-Corning was beating back a hostile takeover attempt, Owens-Illinois, now a privately held company under KKR, was doing its own hostile takeover deal. The KKR takeover in 1986 saddled the company with debt. But in June 1987, O-I sold its Forest Products Division, along with its 750,000 acres of timberland, for $1.5 billion to Great Northern Nekossa Corp. The company finally had some cash.[27] It began expanding its Health Care and Retirement Corporation of America (known formerly as Health Group, Inc.) by building and buying nursing homes, owning 125 facilities in 19 states by July 1987.[28] In September of that year, it began a hostile takeover bid for Brockway, Inc., a glass and plastics manufacturer in Jacksonville, Florida. The company was one of Owens-Illinois's chief competitors, with 14 percent of the domestic market.[29] The takeover was funded by the same banks that funded the takeover of O-I the year before.

But the purchase hit a snag when the Federal Trade Commission sought to block it over concerns that it would decrease competition in the glass container industry. The FTC suggested O-I compromise, and the company delayed its takeover action while it negotiated with the regulatory agency. But the talks fell through, and after 11 delays, O-I decided to go forward with the purchase, stating it would fight the FTC in court. Early in 1988, the U.S. Court of Appeals ruled against the FTC, and O-I went ahead with the purchase. While giving Owens-Illinois a huge share of the glass container business, it cost the company nearly $750 million.[30]

Toledo Pays the Price

The real victims in the rush of corporate takeovers in the glass industry were the people of the city of Toledo. A devastating article in the magazine *Metropolitan* in October 1987 captured the personal price of corporate decisions. Interviews with managers at Owens-Corning and Owens-Illinois described people who were trying to cope. One person from Owens-Corning who was interviewed said that many people saw the signs of layoffs, but were "stuck" in Toledo because of children and spouses. "I have one child at home, a senior in high school. I told him and my wife not to be alarmed, so when my manager took me aside and gave me the news it was a terrible shock. He had difficulty articulating the decision. I felt panicky; I felt real physical pain," he said.[31] Adding to his anger is the fact that he felt the corporate leaders had let down the employees. "The people with six-figure incomes and golden parachutes were getting paid to look out for the company. They should have seen the possibility [of the takeover attempt by Wickes] coming and been prepared for it."[32] Another person interviewed who worked in finance at Owens-Illinois also tried to articulate the personal pain he felt. "Nothing can cushion the blow of being laid off. It knocked me out; I was in a haze for almost a week. . . . The longer things went on, the more haunted I felt."[33] An article in the *Blade* about the layoffs at Owens-Corning described one worker waiting for the phone to ring to be told he was being "de-hired." "Even though I thought I was prepared for the moment, the emotions came through. I didn't know what was going to happen to me or my wife. I had a son in college and a daughter getting married. How would I pay tuition? How would I pay for the wedding? I was scared."[34] Owens-Corning established a Career Transition Center to help laid off employees find other positions. Gary Cuff, who headed the transition office, optimistically predicted that most workers would find similar positions with other companies.

By 1990, signs of the impact of a decade of recession and restructuring were everywhere in Toledo. Corporate takeover fever in the decade of the 1980s saw some Toledo companies disappear completely, such as the Questor Corporation and Sheller-Globe. Champion Spark Plug was purchased by a Houston company, and its last Toledo plant closed its doors in 1991 after 80 years in the city.[35] An article appearing in the *Blade* in January 1990 described 250 people lining up for two hours in the cold before the unemployment office opened to file for benefits.[36] The 1990 federal census showed that Toledo's population had declined by over 6 percent. One positive note was that Owens-Illinois spun off its health care business in 1991 as the Health Care and Retirement Corporation, which was headquartered in

Toledo and traded on the New York Stock Exchange. Some smaller firms were created by those who formerly worked in research and development of the glass corporations.

Toledo vice mayor Carleton Finkbeiner and councilman Jack Ford (both of whom would become mayors of Toledo) sought to recover some of the loss to the city by requesting that Owens-Illinois give up the 20-year tax abatement it had received to build its new world headquarters. The city leaders argued that the terms of the abatement included O-I creating 1300 new jobs, which never materialized—in fact, the workforce had been significantly downsized. The abatement was worth about $1 million a year, money that was desperately needed by Toledo government in 1991. In an editorial, the *Blade* criticized the move to take back the abatement, noting that it was made in 1978, long before the era of corporate takeovers. "We are all older, perhaps wiser, and somewhat sadder now. Former President Ronald Reagan's cowboy capitalism came along, and all of a sudden American companies, surveying their international competition, found it easier to swallow one another in hundreds of not-so-easy payments than to develop new products and sell vigorously against the onslaught of newly industrialized nations of Asia."[37] But, the *Blade* concluded, while O-I's grand plans for the downtown were greatly impacted by the corporate takeover era, the company remained a major player in the city, and the downtown of 1991 was a far better place than it had been 20 years before. In the end, the city council dropped the effort.

Some bright news for Owens-Illinois and the city was announced in 1991, when the company stated its intention to become publicly traded again. It raised the funds go public by selling off perhaps its most important—at least historically—asset, its Libbey Glass division. In June 1993, Libbey Glass became a public company, although the move left Libbey burdened with a large debt.[38] It also had to underwrite the cost of some administrative functions that had previously been provided by Owens-Illinois.

Once public and on its own, Libbey, too, entered the acquisition frenzy. In 1995, it purchased Syracuse China, which produced tableware, largely for the commercial market. It also expanded globally, buying Vitrocrisa, the largest tableware manufacturer in Mexico. It also bought World Tableware, a major importer of products for the commercial food service industry.

The Past Returns to Haunt

One of the reasons Owens-Corning Fiberglas had survived the energy crisis of the 1970s better than the other Toledo glass companies was because of its focus on insulation, a product much in demand by consumers seeking to save on skyrocketing energy costs. So it was ironic that a decision made in 1958 to purchase a line of insulation products from Owens-Illinois would came back to haunt Owens Corning in 2000, and nearly destroy it.

Kaylo was an asbestos-based insulation usually installed around hot pipes such as those in furnaces, air ducts, and boilers. It was also heavily used by the U.S. government in navy ships during and immediately following World War II. Kaylo was named as such because insulation was ranked by the industry by its "K" factor—the lower the "K" value, the better the insulation—hence the name Kaylo. The product was 12–15 percent asbestos mixed with quartz. Articles appearing in medical journals as far back as 1938 began to report on health issues related to asbestos products, particularly the usually fatal lung disease asbestosis (later referred to under the general term mesothelioma).[39] Asbestosis was characterized by the development of tumors in the lungs that eventually robbed the patient of the ability to breathe. In 1943, Owens-Illinois contracted with the Saranac Laboratory to test Kaylo for possible hazards for both workers producing it and the thousands of plumbers, millwrights, and construction workers who installed it.

In his book *Outrageous Misconduct: The Asbestos Industry on Trial*, author Paul Brodeur traced the history of the litigation of asbestos cases. Brodeur cited reports of tests conducted by Dr. Leroy Gardner at the Saranac lab in 1943 that should have caused alarm.[40] In a letter dated February 12, 1943, sent to Kaylo manufacturer Owens-Illinois, Dr. Gardner warned that risks from the product "should be considered from the standpoint of employees working in the plant where the material is made or where it may be sawed to desired dimensions and also considered from the standpoint of applicators or erectors."[41] Dr. Gardner's research was continued at Saranac by Dr. A. J. Vorwald, who became the new director of the lab in 1948. Dr. Vorwald reported to Owens-Illinois in November 1948 on animal tests that showed "unmistakable evidence of asbestosis has developed, showing that Kaylo on inhalation is capable of producing asbestosis and must be regarded as a potentially hazardous material."[42] A final report on Dr. Vorwald's experiments was sent to Owens-Illinois in 1952. That report stated, "The results of the study indicate every precaution should be taken to protect workers against inhaling dust [from Kaylo]."[43] O-I began to give annual x-rays to workers producing Kaylo at its plants in Berlin and Sayreville, New Jersey.

The announcement of the purchase of Kaylo by Owens-Corning Fiberglas in 1958 contained none of this ominous information. Kaylo was heralded in the company's newsletter because its "highly efficient insulating properties of the material combined with its high strength, ease of handling and fabricating on the job [has] won quick acceptance for Kaylo production from insulation contractors and maintenance engineers."[44] Owens-Corning continued to produce Kaylo with asbestos at the Berlin, New Jersey, plant until 1971. When more reports began to link asbestos with lung disease, Owens-Corning and other manufacturers like Johns-Manville put warning labels on its asbestos-based products.

In 1982, Johns-Manville, which produced most of the asbestos-based products sold in the market, was forced to declare bankruptcy because of worker health claims totaling an estimated $2 billion.

Owens-Corning Fiberglas turned to its insurance carriers to settle worker asbestos-related claims in the 1980s. In 1985, the insurance companies that were insuring Owens-Corning along with 34 other companies decided on a novel approach to settling the claims—they pooled their resources into something called the Asbestos Claims Facility (ACF), which was designed as a way to save money and speed up the process of paying victims. An article in the *Wall Street Journal* in 1988 described the intent of the ACF: "As it was first conceived, the Asbestos Claims Facility was supposed to represent a truce among all combatants. Its founders envisioned a one-stop settlement shop for claims, a system that would offer a faster, less costly and more orderly way to resolve claims than in court."[45] But the move did not work. Many victims refused to enter into settlements with the organization, and instead used the settlements paid out by the ACF as grounds for separate suits against individual companies.

In 1988, when as many of 1,300 claims a month began to be filed, it was clear that the concept of a pooled settlement fund was not viable. Some firms that were a part of the ACF fought efforts to settle with claimants, which delayed many from getting timely payments. The ACF was dissolved. Another pooled settlement was attempted in 1990 with the Center for Claims Resolution, but Owens-Corning decided to pull out and manage it own cases when the costs of participating seemed out of line with the number of claims related to Owens-Corning products.

In March 1992, over 8,000 plaintiffs were ready to go to trial against 14 asbestos companies in what was called the largest asbestos trial in history. Owens-Illinois (which despite selling its Kaylo operation to Owens-Corning 34 years before was still being sued) and Owens-Corning (and others) reached a settlement with the plaintiffs on the eve of the trial.[46] But that did not end the nightmare for either company. Between 1990 and 1996, Owens Corning (which in 1992 dropped both the word "Fiberglas" and the hyphen between the two original founding companies from its name) set aside over $2 billion to pay asbestos claims.[47] The number of lawsuits filed against the company rose to 30,000.[48] Most of them were not workers who produced Kaylo, but rather those who installed products that contained asbestos.

In the middle of the asbestos litigation struggle, Owens Corning announced that it was looking to relocate the company from the Fiberglas Tower, the distinctive building that dominated the Toledo skyline that had been built by Harold Boeschenstein in 1967. After One SeaGate, the Fiberglas Tower was the second largest office building in downtown Toledo. The company said that negotiations with the building's owners had stalled, and the skyscraper required major renovations. In 1992, William Boeschenstein retired from the company started by his father, and was replaced by Glen H. Hiner, who came from an executive position at General Electric. In 1997, after much negotiation with the city, the county, the Toledo-Lucas County Port Authority, and the Labor-Management-Citizens Committee that netted incentives of $100 million, Owens Corning under Hiner moved into a brand-new campus-like headquarters in the Middlegrounds area of the downtown.[49] The building was designed by noted international architect Cesar Pelli. To build the headquarters on the site, the city purchased a development of nearly new expensive condominiums located on the riverfront, and sold them to a developer who literally floated them down the Maumee River to a new location. Among those selling their condo in the deal was then-Toledo mayor Carleton Finkbeiner.

Owens Corning and other asbestos defendants began to look to the courts for protection against worker claims. The company noted that much of the money earned in asbestos settlements went not to sick workers, but to highly paid lawyers. One, Robert Motley from South Carolina, who was estimated to have made $200 million handling asbestos lawsuits, named his dog "Kaylo."[50]

Despite the staggering claims against it, in 1997 Owens Corning spent $640 million for the Fibreboard Corporation, a company with its own significant asbestos liabilities.[51] The purchase was intended to achieve the goal of CEO Hiner for Owens Corning to become a major manufacturer of vinyl siding. Another effort at creating a pooled asbestos settlement process was started that year with the creation by Owens Corning of the National Settlement Program (NSP) to settle 176,000 cases. Within a year, the number of cases pending with the NSP was up to 237,000, and more plaintiffs continued to sue Owens Corning outside of the settlement program.[52] The company stopped actively fighting the lawsuits in hopes of getting to the end of the claims for a product it had stopped making in 1971. In 1998, Owens Corning agreed to pay another $1.2 billion to settle suits without lawsuits for two years.[53]

By August 2000, rumors were rampant that Owens Corning would have to declare bankruptcy in order to get out from under the asbestos lawsuits. The company denied this, but its bond rating by Standard & Poors dropped to "junk" status, and company's stock, which once sold for $82 a share during the frenzy created by Wickes takeover bid in 1986, dropped to $9.25.[54] One of the worries was that the trust fund that had been established to pay asbestos claims against the recently acquired Fibreboard Corporation was not large enough to cover all of those claims.

Within two months, the company—which had already paid out 440,000 claims and now faced another $3 billion in claims against it—voluntarily filed for Chapter 11 bankruptcy protection. Ironically, the company had only made $135 million from the sale of Kaylo insulation over the entire course of its production.[55] Owens Corning was one of at least 25 companies that once made asbestos products to be forced into bankruptcy. The sale of shares of company stock, which were selling for less than $2 a share, was suspended by the New York Stock Exchange with the announcement of the filing. Within a day, the price of the stock dropped to less than a dollar a share. Employees, who owned 12 percent of the company's stock, lost the most—some of them their entire life savings because of the bankruptcy. Many of the losers were older, and depended on their stock to help fund their retirement. In its filing, Owens Corning listed $7 billion in assets, and $5.7 billion in debt.[56] The bankruptcy was not only the largest ever in the history of Toledo, but also one of the largest in the country up to that time.

Glen Hiner, who had once been hailed by *Business Week* magazine as one of the six best corporate managers in the country, was suddenly criticized for some of his decisions, such as the acquisition of Fibreboard Corporation and its $1.8 billion in asbestos liabilities. Once again, the company turned to Washington in hopes of getting some legislative relief. The company stated that the government should help fund the asbestos payouts because it had required the use of asbestos-based insulation in navy warships.[57] That lobbying effort was short lived.

Three months after declaring bankruptcy, Owens Corning put forth a $70 million bonus program to reward top managers who stayed with the company throughout the bankruptcy.[58] Glen Hiner was not one of them—in early 2002, he retired from the company and was replaced by David Brown as chief executive officer. Along with chairman and chief financial officer

Michael Thaman, the two worked to develop a plan to exit bankruptcy. In 2002, the company posted a $2.8 billion year-end loss.[59] Bankruptcy documents showed the company's debt to be much more than the $5.7 billion originally estimated, with the actual debt somewhere between $10 billion and $12 billion.[60] Creditors were especially angry about the purchase of Fibreboard.

In January 2003, Owens Corning submitted a reorganization plan that would finally allow the company to exit from bankruptcy. The move kept creditors from filing their own plan for the company. When the plan was filed, the company's stock was selling for 60 cents a share.[61]

The company's financial picture seemed to be symbolized by its still vacant Fiberglas Tower, the beacon of company dominance that had become little more than a 28-story antennae tower. Ironically, it was asbestos insulation in the building that had kept it vacant because no developer was willing to foot the bill to remove it. In February 2003, water pipes in the building burst, and water began cascading down the building from the former executive suits on the top floor. It quickly froze, and ice encased the structure. At the time, the building was owned by two brothers from Lansing, Michigan, who had purchased it as tax shelter.

But that was not the end of Owens Corning's bad news for Toledo. The bankruptcy filing forced the company in March 2003 to try to get out of its lease on its new headquarters building along the river. That $100 million structure had been built with $85 million in bonds sold by the Toledo-Lucas County Port Authority along with a loan from the state of Ohio. The company said it could no longer afford its annual lease payments of $30 a square foot, and demanded that the rent be reduced or it would look for a new headquarters elsewhere, perhaps outside of Toledo.[62] After much negotiation, the Port Authority agreed to give up $18 million in rent, and gave the company a reduced rental rate.[63]

In Search of Greener Suburban Pastures

As the original manufacturer of Kaylo, Owens-Illinois also faced litigation over the product. Like Owens Corning, Owens-Illinois fought with its insurance companies over claims to cover the costs of the settlements. In 1995, one of its largest insurance companies agreed to pay $140 million to those filing claims. The CEO, Joseph Lemieux, expressed relief that the asbestos issue might be behind them. "These settlements mark the beginning of a new chapter in the history of Owens-Illinois," Lemieux stated.[64] But as with Owens Corning, the lawsuits kept coming.

In 2000, the company announced it was eliminating about 500 jobs in Toledo because of rising energy costs and asbestos-related litigation, a move that cost it $100 million in severance payments. The company had already paid out $1 billion to settle asbestos cases, and was facing an additional 14,000 new cases. Some 7,000 new claims were being filed each year.[65] But while Owens Corning had to declare bankruptcy that year to get out from under its asbestos lawsuits, the executives of Owens-Illinois strongly dismissed any conjecture by the business community and media of similar action by O-I. The company felt it could better manage the liability because it had stopped selling Kaylo in 1958, and the number of possible cases of illness as a result of the product during the time it was produced by O-I was declining.

But when Owens Corning declared bankruptcy, the move impacted the stock price of Owens-Illinois, which plummeted to less than $3 a share in December 2000. Like Owens Corning, O-I was dropped from the Standard & Poor's 500 Index. In April 2001, the company received a $4.5 billion line of credit from a syndicate of banks that was vital to keep the company out of bankruptcy.[66] The following month, CEO Lemieux assured shareholders at the company's annual meeting that company had turned a corner, and that the number of asbestos claims was sure to decline.

Also in 2001, Edwin Dodd, who led Owens-Illinois when it spearheaded the efforts to redevelop downtown Toledo in the 1980s, died. The *Blade* noted his passing—and the passing of an era of corporate leadership—in an editorial. "In a high-tech world surrendering to corporate downsizing, capricious relocations, and misplaced loyalties, Ed Dodd was an anomaly, a man who resisted all overtures to move O-I out of Toledo, a man who believed that a Fortune 500 company should give, not just take, from the community it calls home," the editorial stated.[67]

It was as if the *Blade* had a premonition about the company's future in downtown Toledo. Joseph Lemieux resigned as the head of O-I in 2003, and in 2004 the company brought in an outsider, Steven McCracken from DuPont, to succeed him. In less than a year, McCracken began to hint broadly that the company was considering exiting from downtown Toledo. The lease on One SeaGate was set to expire in late 2006, and if the company was to find a new headquarters elsewhere, the decision had to be made quickly.

Immediately, Toledo officials developed an incentive package worth an estimated $8 million to try to keep Owens-Illinois downtown.[68] It included money for renovating One SeaGate. But competing with the downtown location was the large tract of over 400 acres of land in Perrysburg (about 10 miles south of Toledo) that the company had purchased in 1966. It had three production facilities there, and one of the buildings was empty after O-I sold its plastic bottle production facility to Graham Packaging Co., LP, in 2005. The area around the Perrysburg plants had convenient access to interstate highways, and had recently been developed into an upscale shopping mall and residential area known as Levis Commons, named for William and Preston Levis. Owens-Illinois spokesperson Sara Thies said the company was "in the midst of a study to consider several options, including staying downtown, moving to Levis Park in Perrysburg, or moving out of state. Our decision won't be based solely on what's financially best, but also what best meets our cultural needs," she said.[69]

Like Owens Corning, Owens-Illinois's leadership expressed a desire to move away from the hierarchical management model that a skyscraper encouraged and transform into a more casual, horizontal management style. In addition to this change in corporate culture, another reason for the move was that O-I occupied little of its 32-story signature building downtown. Only four floors were then being used by the company, yet because of the building's leasing agreement, it was paying $13.5 million a year in rent.[70]

In an interview in early February 2005, McCracken said there was a 60 percent chance the company would move out of the downtown.[71] In addition to Perrysburg, other options considered by the company included New York, Miami, Detroit, and London. The *Financial Times of London* fueled the speculation about a move to England to create a new global headquarters there. The company quickly denied the rumor.

In a presentation to the Owens-Illinois retirees, McCracken laid out the arguments for and against leaving downtown Toledo. While he admitted that the company's heritage was in Toledo and the One SeaGate building was symbolic of that long history, he said the decision had to be made based on the future, not the past. Because the company already owned the land in Perrysburg, it was the most cost-effective solution. Employees who were surveyed about where they wanted the new headquarters located seemed to favor Perrysburg, which was closer to where many of them lived. The retirees, some of whom had been instrumental in planning One SeaGate, were less enthusiastic. "They have a strong feeling about O-I's heritage and the importance of the headquarters being in the city where the company was born," said Jack Paquette, former vice president and assistant to Edwin Dodd.[72]

By May, the obvious was announced. The company would move from its gleaming, 24-year-old skyscraper, and would build a new campus-like headquarters on its land in Perrysburg. The decision was helped by a roughly $6 million an incentive package offered by Perrysburg. In addition, Ohio's Department of Development offered incentives of $11 million to keep the company in the state.[73]

With the impending exodus from One SeaGate, Toledo city officials began their search for new tenants for the building. About a quarter of the building was vacant, even before O-I began to move out. The building was then owned by Newkirk Master Limited Partnership of Boston, and the mortgage was held by the RVI Group of Stamford, Connecticut. It looked as though RVI would take over the building, as the owners indicated they would default on the mortgage because they were unable to make a $32 million final balloon payment.[74] The building constructed in 1982 for $100 million was valued for tax purposes in 2005 at $41 million, but realtors put the value even lower, at between $20 and $30 million.[75] In the end, an agreement was reached with Fifth Third Bank of Northwestern Ohio to occupy parts of the building. The company did not occupy the opulent top floors where O-I's corporate leadership had been housed. With their curving staircases, gold-colored bathroom fixtures, and huge chandeliers, these floors remained empty.

The new corporate headquarters building for Owens-Illinois was a three-story building of about 100,000 square feet that cost $20 million and was designed by Albert Kahn Associates of Detroit—one-eighth the size of One SeaGate. The space was much less lavish, and instead focused on functionality and efficiency. Located in the midst of the Levis Commons shopping area, it required few of the employee amenities that had been a part of the downtown facility. The building did, of course, feature large expanses of glass on the exterior.

Owens-Illinois opened its new headquarters in August 2006. To honor its past, the road leading to the building was renamed One Michael Owens Way. One person not present at the opening of the new facility was president Steven McCracken. The month before, he had undergone surgery for stomach cancer. A company spokesperson said it was hoped he would return to O-I leadership within weeks, but by November, he was forced to step down because of his illness. He was replaced by a member of O-I's board of directors, Albert Stroucken, the CEO and chairman of H. B. Fuller of St. Paul, Minnesota. In February 2008, McCracken died from complications of cancer. He was 54 years old.

Libbey, Owens, and Ford Become History

When the British firm Pilkington PLC took over Libbey-Owens-Ford in 1986, it indicated it would retain the company's name because of its strong national identity and its tie to the three historical figures of the glass industry.[76] But in May 2000, Pilkington reversed course. The British firm said its name, not Libbey-Owens-Ford, was more appropriate for a business with a global presence. On May 27 of that year, in a ceremony at the company's headquarters in downtown Toledo, the company raised a new company banner proclaiming the Pilkington name. Among those unhappy with the decision was Toledo mayor Carleton Finkbeiner. In a statement, the mayor said, "I don't like it at all. The history of Libbey-Owens-Ford is rich and deep. Edward Drummond Libbey, Michael Owens, and Edward Ford built L-O-F, not Pilkington. Shame on Pilkington."[77] The *Blade*, too, lamented the passing, noting that the company's founders had been the most significant figures in Toledo's history.

Pilkington countered with a full-page advertisement in the newspaper headlined "One Pilkington." "From Pilkington Libbey-Owens-Ford to Pilkington, the change is not made without a deep sense of gratitude for Libbey-Owens-Ford's founders and for the thousands of people who helped grow the company's brand over its 70-year heritage. What's past is not forgotten. But, today, our vision must be focused on the future. It is one big world out there and one Pilkington will serve it best."[78]

At the time the company changed its name, the workforce at the Rossford L-O-F facility, which had once numbered 8,000, was down to 325.[79] Many of those workers were close to retirement age. Warren Knowlton, president of Pilkington Automotive Worldwide and president of Pilkington North American operations, said the firm was committed to keeping the Rossford production facility, and promised an investment of $75 million to rebuild its float glass lines. Still, the *Blade* could not help noting that the name change meant that a part of Toledo had died.[80]

Within two years of the firm's name change, the company began a restructuring of its North American operations. As part of that restructuring, about 275 employees accepted retirement, and the company closed the East Toledo technical center. Warren Knowlton was also out of a job as head of North American facilities. There was talk that the headquarters would move to North Carolina. Profits that year declined by 4 percent. Rising energy costs were the reason cited by the company for the dip in profits.

In 1993, Pilkington had sold its headquarters building on Madison

Avenue—once the pride of L-O-F—to the city of Toledo for $5.2 million.[81] The city paid for the building through loans and grants from the state of Ohio. Toledo hoped that by purchasing the building, it could encourage the company to remain there. Pilkington agreed to pay rent on the building of $50,000 a month.[82] The city tried unsuccessfully for two years to sell the building before developer David Ball and his Ohio Building Company purchased it in 2002 for $2.2 million after Pilkington refused to bid on the property or commit to a long-term lease. Ball sought a low-interest loan from the Ohio Department of Development and also a $5 million capital improvements loan from the state to finance the purchase and modernize the building.[83]

With its lease set to expire on December 31, 2003, Pilkington refused to say whether or not it would continue to maintain its North American headquarters in the building, or even if it would stay in Toledo. Much of the building's office space was taken over by the Hylant Group, a local insurance firm.

Through negotiations with the administration of Toledo mayor Jack Ford, the company agreed in June 2003 to stay—but at a cost. Under a new "downtown employment incentive program," the city agreed to pay Pilkington and the Hylant Group each $75,000 a year for 10 years as a rebate on their income taxes in order to retain the jobs in the city.[84] At the time of the agreement, Pilkington had 166 employees in the building, and they were housed on just four floors.

And there was yet one more shake-up to occur in the glass industry in Toledo in 2003. The American Flint Glass Workers Union, founded in 1878 and headquartered in Toledo since 1904, decided to merge with the United Steelworkers of America. The union, which had been plagued by declining membership due to layoffs in the glass industry and financial problems, saw the merger as a means for survival. The move brought the 12,000 AFGWU members into a union organization of 450,000, and one that already included the Aluminum, Brick & Glass Workers Union. While the union's building on Byrne Road in Toledo was to remain in union hands, it was still a stunning change for the organization. When the membership approved the merger in June 2003, *Blade* reporter Gary Pakulski noted, "Toledo's claim to be the glass capital of the world appeared close to losing another chip last night."[85]

By the end of the tumultuous year of 2003, the city of Toledo had managed to maintain two of its glass company headquarters in the downtown. In addition, its national union of glassworkers too continued a presence in the city, although as merely a part of a much larger entity. Each of these changes seemed to signal the glass industry's past in the city was perhaps more glorious than its future.

The Museum That Glass Built

While the glass industry faded from the forefront of Toledo's economy, the importance of the art museum created by the founder of the industry in the city, Edward Drummond Libbey, continued to expand, both in terms of its collections and its reputation. And that reputation was enhanced by its collection of art glass and its support of glass artistry.

After the death of director George Stevens in 1926, Blake-More Godwin, who had begun his career as a curator in 1916 and had worked with both Stevens and Libbey, became the Toledo Museum of Art's director. Godwin, in turn, hired Otto Wittmann Jr. in 1946 as associate director. Wittmann was a graduate of Harvard University, had worked as a curator at what is today the Nelson-Atkins Museum of Art in Kansas City, and had taught art history. During World War II, Wittmann worked for the government's Art Looting Investigation Unit, tracking down art stolen from European Jews by the Nazis. While pursuing stolen artwork, Wittmann gained extensive knowledge of both the art and museums of Europe, knowledge that he brought with him to Toledo.

In 1959, Wittmann succeeded Godwin as director of the Toledo Museum of Art, and in this position, he sought to expand the museum's collections and its reputation internationally. Under his direction, what was once a respectable museum became world-renowned. The money to expand the museum's collections came from the trusts set up by Edward Drummond Libbey and his wife, Florence. While substantial when they were first established, the trusts continued to grow in value. In 1928, $239,000 was given to the museum from the E. D. Libbey trust. By 1975, the amount given to the museum that year had grown to $1.9 million.[86] Similar growth was seen in the value of the Florence Scott Libbey trust, with the museum's share increasing from $12,500 in 1946 to $415,000 in 1975.[87]

Because of the museum's connections to Libbey, director Wittmann continued to develop its glass collection. Wittmann also encouraged the development of a whole new movement in glass artistry that became known as the "studio glass movement." In 1962, he invited Harvey Littleton, a ceramics artist on the faculty of the University of Wisconsin–Madison, to present an experimental workshop on glassblowing.[88] Littleton was from a glassmaking family from Corning, New York, and had studied art at the Cranbrook Academy of Art near Detroit. Prior to this workshop, glass artists produced their works using industrial glass furnaces, which limited experimentation and artistic expression. From March 23 to April 1, 1962, Littleton presented his workshop to seven students using a small handmade glass furnace built in a garage on the grounds of the museum.

Despite his experience with the medium of glass, Littleton could not get the small furnace to work properly. The glass would not melt correctly, and he lacked experience in the techniques of glassblowing. He invited his friend, Dominick Labino, to help solve some of the technical problems. Labino was vice president and director of research at Johns-Manville, the company formed from the early fiberglass manufacturing efforts at Libbey-Owens-Ford, which L-O-F spun off in 1954. Because fiberglass was produced from marbles, Labino convinced Johns-Manville to donate some to the workshop. He also rebuilt the furnace with special bricks that could withstand the intense heat.

Labino was a talented researcher and inventor, and held 60 patents for various glass products. One of these was for pure silica fiber, a product used in the insulating tiles that protected the outside of the Apollo and Mercury spacecrafts and the Space Shuttle from burning up upon reentry into the earth's atmosphere. But he was more than a technician and engineer—Labino was also an artist who had grown frustrated working for industrial glass companies. He not only shaped the studio glass movement through his technological improvements, but he also became a well-regarded glass artist. The third person who was called into the workshop was Harvey Leafgreen, a retired glassblower from Libbey Glass. Leafgreen instructed the participants on the art of blowing glass.

With the success of the first workshop, Littleton organized a second one in June 1962. He spread the idea of studio glass to others, and gained a national reputation as the leader of the movement when he appeared in an extensive spread in *Life* magazine. In 1971, he published a landmark book on studio glass titled *Glassblowing: A Search for Form.*

Otto Wittmann and the Toledo Museum of Art supported the continued development of the studio glass movement. In 1969, the museum built the Glass Crafts Building as the first museum-based glass studio. The building was funded through donations from Libbey-Owens-Ford, Owens-Illinois, and Owens-Corning Fiberglas.

Harold Boeschenstein organized the first fund-raising campaign for the Toledo Museum of Art that year. Through donations both given and raised by Boeschenstein, and a major donation from William Levis, in 1970 the museum opened the Art in Glass Gallery. The modernistic gallery gave the museum a dedicated space devoted to its expansive glass collection. Among the most popular pieces displayed in the new gallery was the cut glass punch bowl that had been made by Libbey Glass for the 1904 St. Louis World's Fair. But the gallery also included examples of everything from ancient glass to pieces produced by the new glass artists of the studio glass movement. One large piece was a mural of glass panels by Dominick Labino, whose work was becoming internationally known. *Vitrana*, as the piece was called, was made by Labino casting molten glass into molds and laying in bright colors. It greeted visitors as they entered the gallery.

Otto Wittmann served as director of the Toledo Museum of Art until 1976. He was succeeded by Roger Mandel (1977), David Steadman (1989), Roger Berkowitz (1999), Don Bacigalupi (2003), and Brian Kennedy (2010). Under Berkowitz, the museum began planning for a new building to be devoted entirely to its glass collection. The name chosen was the Glass Pavilion, which harkened back to Edward Drummond Libbey's exhibit hall at the 1893 World's Columbian Exposition.

The two architects selected for the project were from Japan, and had never had a commission in the United States. Ryue Nishizawa and Kayuyo Sejima of SANNA Ltd. were commissioned to create a signature building to be built directly across Monroe Street from the museum. The Glass Pavilion was to be as significant in its design as the original museum building designed by Edward B. Green and Harry W. Wachter in 1912 and the Center for the Visual Arts designed in 1993 by internationally known architect Frank Gehry.

The Glass Pavilion met the challenge. When it opened in August 2006, it was a jewel—transparent and encased nearly entirely in glass. It was as much a piece of glass art as the objects displayed within it. Many of the pieces from the Art in Glass Gallery were moved to the new building, along with several important new acquisitions. One of the new additions was a nine-foot-tall chandelier by Seattle artist Dale Chihuly. In addition to the museum's glass collection, the Pavilion included a working glass studio with furnaces that could be seen glowing at night. Glass artists gave demonstrations of their work to the public as well as workshops and classes to aspiring glass artists.

The Glass Pavilion was not only world-class architecture, but its construction also reflected the globalization of the glass industry. In addition to being designed by a Japanese firm, the large formed-glass walls were produced in Shenzhan, China; the lighting consultants were from London; the heating and air conditioning system came from Austria; and the hot-glass studio was designed in New York.[89] The glass panels could not be made in the United States—even by Pilkington—because no U.S. firm could meet the architect's specifications. The 1,300 pound bent, laminated panels used techniques that had originally been developed by Libbey-Owens-Ford in Toledo, but they had to be produced in China by the Avic Sanxin Co., a firm that specialized in making custom glass.[90] At the time of its production, the United States had only 33 float glass lines in production—with two in Toledo—and these manufactured only standard, high-volume products. China, on the other hand, had over 150 float glass lines in production, and Avic Sanxin invested $500,000 in equipment to make the custom glass required for the Glass Pavilion. Some in Toledo lamented the lack of local firms involved in the project.[91]

With the success of the Glass Pavilion (the architects won the Pritzker Prize in 2010 for their designs, including the Glass Pavilion), local history enthusiasts sought to purchase and renovate the former home of Edward Drummond and Florence Scott Libbey that was located on Scottwood Avenue, adjacent to the Glass Pavilion. The group began a fund-raising campaign to raise $3.3 million to construct a museum about Toledo's glass heritage, a subject not covered in any great extent in the Glass Pavilion.[92] This project continues today. While the house has been purchased and stabilized, the museum has not been completed.

A "Bright" Future for Glass?

At the time of the opening of the Glass Pavilion, the *Wall Street Journal* carried an article entitled "In Toledo, the 'Glass City,' New Label: Made in China." The article noted that China produced 45 percent of the world's glass, producing enough every 15 minutes to cover a 100-story skyscraper.[93] The country's success was the result of lax environmental and worker safety regulations and cheaper labor and energy costs.

The one "bright" spot in Toledo's glass industry in the first decade of the 21st century was solar energy. Both Owens-Illinois and Libbey-Owens-Ford had experimented with producing solar collectors during the energy crisis of the 1970s, but neither firm developed a profitable product line. However, not only did a smaller firm located just outside Toledo in Perrysburg succeed in the solar energy market, but its founder is often referred to as the "father" of commercial-scale solar energy.

Harold McMaster was hired in 1940 as an engineer by Libbey-Owens-Ford. Working on projects for the war, he invented a rear-vision periscope and a method for de-icing airplanes. In 1942, he became manager of L-O-F's optical glass laboratory, and developed advances in windshield production.

In 1948, McMaster left the company to start his own firm called Permaglass, which made glass for appliances and television sets. In 1971, McMaster, Norman Nitschke, and Frank Larimer founded a new company named Glasstech. The company manufactured a machine developed by McMaster that made large panels of strong, clear tempered glass. The machine was so successful that it was sold all over the world, with an estimated 80 percent of automotive glass and 50 percent of architectural glass produced using the Glasstech machine.[94]

During the energy crisis of the 1970s, McMaster began to consider how his glassmaking machinery could be used to produce solar cells. Norman Nitschke recalled, "We got interested in photovoltaics [a form of solar energy collection] because it was another outlet for glass from our end. If you look at the fields of solar panels that are beginning to dot the landscape worldwide, that's a lot of glass."[95] In 1984, Nitschke and McMaster founded Glasstech Solar, Inc., which utilized Glasstech glass as the foundation for photovoltaic cells. Rather than make solar panels for small installations such as on homes and buildings, the company focused on large rows of solar collectors that could feed into the electric grid. But the company struggled to make its amorphous silicon coatings strong enough to survive extremes of weather.

In 1987, McMaster and Nitschke created a new company, Solar Cells, Inc., which focused on research involving cadmium telluride as a semiconductor for solar energy panels. Solar Cells rented production space at the University of Toledo, and researchers from UT received grants from local, state, and federal agencies to assist with developing the technology. In 1998, the company developed the Vapor Transport Deposition technology that deposited cadmium telluride on large glass panels—a key breakthrough in thin-film solar collection.[96] To provide the company with needed capital to continue its research, in 1999 Solar Cells merged with True North Partners, LLC, of Phoenix to form First Solar, Inc. Sadly, the company's founder, Harold McMaster, died in August 2003. While First Solar continued to operate a factory in Perrysburg, in 2010 the company moved its headquarters to Tempe, Arizona. It remains as one of the largest manufacturers of thin-film solar cells in the world.

With the dramatic reduction in the glass industry workforce in Toledo (in 2010 it was estimated at just 2,500 workers), city fathers embraced solar and alternative energy as the future for the city's economy.[97] In the midst of the recession of 2009, when unemployment rates in the city hit 12 percent, Toledo mayor Carleton Finkbeiner saw solar energy as the natural successor to the city's history as the Glass Capital of the World. "We have about 6000 people at the moment employed in 15 research and manufacturing institutions that are focused entirely on solar energy. We would like to see over a decade that number grow from 6000 to 20,000," Finkbeiner said.[98]

But the transition from "Glass Capital of the World" to "Solar Energy Capital of the World" ran into difficulties. By 2012, the solar energy business worldwide was in crisis. The industry had been subsidized by state and federal governments during the recession of 2008–2009 under the belief that America's future could be secured if manufacturing moved into "green" energy solutions. Countries of the European Union also significantly subsidized their solar energy factories. This led many to the impression that the solar energy industry was financially sound and a good investment.

Several factors undercut the industry. First, as the immediate struggles of the recession began to subside in the United States and economic stimulus money from the federal government dried up, companies that had depended on such funding found it difficult to survive without it.[99] The European Union, weathering its own economic recession, stopped subsidizing its solar energy industry as well. At the same time, the market was flooded with cheaper solar energy products produced by China that, without government assistance, American and European companies could not compete against.

In Toledo, the great solar energy boom has largely faded. The Willard & Kelsey Solar Group, a company founded by several individuals who started in the solar energy field under Harold McMaster, struggled in 2012 to pay back a $10.3 million loan it had received from the state of Ohio.[100] The company had promised to create 450 jobs, but only 80 employees worked for the company, and half of these jobs were eliminated that year. Xunlight, a company founded by University of Toledo professor Dr. Xunming Deng, also struggled to pay back state loans. The company had received a $4 million loan from the Ohio Air Quality Development Authority, $4 million from the Ohio Enterprise Bond Fund, an Ohio Department of Development job creation tax credit of 55 percent for seven years, a $290,000 job training grant, and nearly $7 million in grants from Ohio's Third Frontier program.[101] The company had promised to create 181 jobs, but just over 100 people were on the payroll in 2013, up from a low of 40.[102] The company was moving forward with plans to sell both its flexible solar panels and licenses for its machinery to produce them. In 2012, Dr. Deng left the Toledo company to take over Xunlight Asia.

The Future of the Future Great City of the World

After 125 years in Toledo, in 2013 the future of the glass industry in the city appeared uncertain. The major shake-ups of the companies has subsided—the last takeover was in 2006 when Pilkington Group, which had purchased Libbey-Owens-Ford in 1986, was itself bought out. Ironically, the new owners were Nippon Sheet Glass Co. Ltd. of Japan, the very company that John Biggers of L-O-F had helped to create in the 1920s and had saved after World War II. Nippon Sheet Glass paid $3 billion for Pilkington.[103]

Owens-Illinois has concentrated its production on glass bottles, selling its plastics lines and paying down its debt with the proceeds. In 2010, its Venezuelan facilities, which had already caused problems for the company in 1976 when executive William Niehous was kidnapped there and held in the jungle for three years by rebels, were nationalized by the country's president, Hugo Chavez. The leftist leader expropriated all company assets, a move that surprised O-I executives. Owens-Illinois had to write off $329 million that year because of the loss of the Venezuelan factories.[104] In addition, the company's extensive European operations were adversely affected by the economic downturn on that continent.

Libbey Inc. hired a new CEO in 2011—Stephanie Streeter, the first female executive officer in the company's history. Like O-I, Libbey has recently concentrated on reducing its debt. In 2012, the company signed a 14-year lease on its headquarters offices in the Toledo Edison Plaza in downtown Toledo, thus ensuring that Libbey would remain downtown until at least 2027. It also extended its lease on its popular outlet store in the Erie Street Market downtown.

Owens Corning emerged from bankruptcy a changed company, but one that had finally escaped its asbestos liability. It continued its focus on building materials and Fiberglas-reinforced products.

The only Fortune 500 company still located in downtown Toledo, Owens Corning in 2013 requested financial incentives from the city and the Toledo–Lucas County Port Authority to remain in its headquarters building after its lease expired in 2015. In exchange for the company extending its lease for 15 years, the city of Toledo agreed to extend a tax-increment financing structure until 2024 at a cost to the city of about $7 million. In addition, the Toledo–Lucas County Port Authority agreed to renegotiate the payoff of the headquarters building lease, issuing bonds of up to $8 million to refinance the "tail" of the lease. The company agreed to add 50 jobs to the downtown workforce, and begin making payments of about $400,000 a year to Toledo

Public Schools beginning in 2018 in lieu of taxes owed to the system. The incentive package was negotiated at a time of much publicity about Owens Corning's 75th-anniversary celebration. According to the *Blade*, the company had earned $71 million in profits in the first two quarters of 2013, after a loss of $19 million in 2012.[105]

While the city's glass industry has settled down from the roller coaster of the past three decades, today it has a much diminished presence in the Glass City. And the symbols of that break from the past are many. In 2013, over the protests of alumni and historic preservationists, Toledo Public Schools demolished the Edward Drummond Libbey High School that had been built in 1923. With declining enrollments and shifting populations, the school system no longer needed a large high school on the city's south side. The Fiberglas Tower, once a beacon of the dominance of the glass industry in the city, remained empty. One SeaGate is now emblazoned with the name of its new primary tenant, Fifth Third Bank. The American Flint Glass Workers Union, now the Flint/Glass Industry Conference of the Steelworkers, decided in 2012 to move out of its headquarters building in Toledo to Pittsburgh. And land once owned by Libbey-Owens-Ford is now the home of a new casino, an irony since Edward Ford refused to allow even a saloon to be built in Rossford when he was alive. Today, the old factory buildings of the former L-O-F stand in contrast to the shiny new Hollywood-themed casino next door.

Today, visitors to Toledo can best appreciate the impact of glass on the city by visiting the Glass Pavilion at the Toledo Museum of Art. The industry that Mr. Libbey started is enshrined there, not only in the works displayed within its walls, but by the museum's very presence, thanks to its benefactor. Admission to the museum and the Pavilion also remains free thanks to Mr. Libbey's generosity. Glass may no longer be King in Toledo, as was declared in 1888, but there is no doubt that the industry has shaped the city in complex, profound, and lasting ways.

Notes

INTRODUCTION

1. "Glass is King," *Toledo Daily Blade*, February 18, 1888.

2. James T. Areddy, "In Toledo, the 'Glass City,' New Label: Made in China," *Wall Street Journal*, August 29, 2010.

CHAPTER 1

1. Daniel Drake, *A Systematic Treatise, Historical, Etiological, and Practical, on the Principal Diseases of the Interior Valley of North America* (Cincinnati, OH: Winthrop B. Smith & Co., 1850), 365.

2. Annals of the Mother-House of the Montreal Grey Nuns, as quoted by Edward Ockuly, "History of St. Vincent Medical Center," in *History of Medical Practice in Toledo and the Maumee Valley Area, 1600–1990*, Walter Hartung, ed. (place and date of publication not identified), 26.

3. Clark Waggoner, *History of the City of Toledo and Lucas County, Ohio, pt. 1* (Toledo, OH: Munsell & Company, 1888), 52.

4. Ibid., 377.

5. Harvey Scribner, *Memoirs of Lucas County and the City of Toledo* (Madison, WI: Western Historical Association, 1910), 217–18.

6. S. S. Knabenshue, "The Beginnings of Toledo, 1817, 1832, 1833," *Toledo Blade*, December 12, 1903, 7.

7. For a complete history of the Toledo War, see Don Faber, *The Toledo War: The First Michigan-Ohio Rivalry* (Ann Arbor: University of Michigan Press, 2008).

8. General Joseph Brown, as quoted in Waggoner, 294.

9. *Senate of the United States, Mr. Clayton Made the Following Report . . .* (Washington, DC: Gales & Seaton, 1836) details the terms of the settlement of the Ohio-Michigan border dispute.

10. As printed in Waggoner, 312.

11. For a map of early plans for Ohio's canals, see James Hamilton Young, *The Tourist's Pocket Map of the State of Ohio: Exhibiting its Internal Improvements, Roads, Distances, &c* (Philadelphia: S. Augustus Mitchell, 1835).

12. Jesup W. Scott, as quoted in Scribner, 218.

13. Waggoner, 472.

14. Ibid., 736–37.

15. "Health of Toledo. Aggregate of Deaths by Cholera," *Toledo Daily Blade*, June 30, 1849. The *Toledo Daily Blade* was founded in 1835, and continues to be published today as the *Toledo Blade*.

16. Waggoner, 480, 482.

17. Ibid., 795.

18. As quoted in ibid., 85.

19. Jesup W. Scott, *A Presentation of Causes Tending to Fix the Position of the Future Great City of the World in the Central Plain of North America*, centennial reprint (Toledo, OH: Blade Printing & Paper Co., 1937).

20. Ibid., 5.

21. Ibid., 41.

22. Articles of Incorporation of the Toledo University of Arts and Trades, October 12, 1872, University of Toledo Archives, University of Toledo.

23. Waggoner, 801.

24. Ibid., 802.

25. "Oil and Gas—Toledo and Lucas County," *Toledo Daily Blade*, February 5, 1887.

26. As quoted in Randolph C. Downes, *Industrial Beginnings: Lucas County Historical Series*, vol. 4 (Toledo, OH: Toledo Printing Company, 1954), 49.

27. "Toledo—The New Industrial Center of the West," *Toledo Daily Blade*, May 5, 1887.

28. Ibid.

29. Ibid.

30. "Toledo Rejoices—The Grand Gas Celebration," *Toledo Daily Blade*, September 8, 1887.

31. Waggoner's history contains numerous biographies of the leaders of Toledo in the nineteenth century. A listing of these can be found in the volume on pages xi–xii.

32. "The Garden of Eden," *Toledo Daily Blade*, January 12, 1888.

CHAPTER 2

1. For a summary of the early development of American glass, see Warren C. Scoville, *Revolution in Glassmaking* (Cambridge: Harvard University Press, 1948), 3–10.

2. Ibid., 56.

3. Ibid., 44.

4. Lura Woodside Watkins, *Cambridge Glass, 1818 to 1888: The Story of the New England Glass Company* (New York: Bramhall House, 1930), 3.

5. Details about Deming Jarves and the early history of glassmaking in the United States differ from source to source. Watkins states the date of the construction of the factory that would become the New England Glass Company as 1813. Scoville states it as 1814. Scoville says that Jarves and others bought the factory and property at a public sale in 1817. Watkins states it was purchased by a group of people that included Jarves

in 1818. In some accounts, even the spelling of Jarves's name is different, with at least one reference to Demming Jarvis.

6. Watkins, 11.

7. Scoville, 18.

8. Watkins, 15.

9. Ibid., 18.

10. Scoville, 89.

11. Watkins, 29.

12. Carl U. Fauster, *Libbey Glass since 1818: Pictorial History and Collector's Guide* (Toledo, OH: Len Beach Press, 1979), 5.

13. Scoville, 38.

14. Watkins, 34.

15. Ibid., 157; Scoville, 32.

16. Edward Drummond Libbey's batch book. In the archives of the Toledo Museum of Art.

17. Watkins, 150.

18. Scoville, 90.

19. Fauster, 25.

20. Advertisement in *Pottery and Glass Reporter*, October 13, 1887.

21. "New Glass Works," *Toledo Daily Blade*, April 15, 1887.

22. "Important Meeting," *Toledo Daily Blade*, September 27, 1887.

23. "Going for Another Glass Works," *Toledo Daily Blade*, December 27, 1887.

24. "New England Glass Factory," *Toledo Daily Blade*, January 4, 1888.

25. "The Libbey Glass Works," *Toledo Daily Blade*, January 28, 1888.

26. "The Libbey Glass Works," *Toledo Daily Blade*, February 7, 1888.

27. "Glass is King," *Toledo Daily Blade*, February 18, 1888.

28. "Mr. E. D. Libbey Speaks," *Toledo Daily Blade*, March 31, 1888.

29. "Great is Glassboro," *Toledo Daily Blade*, April 17, 1888.

30. Ibid.

31. "Boss-Town!" *Toledo Daily Blade*, August 18, 1888.

32. Ibid.

33. Ibid.

34. John Staiger's recollections about the arrival of the New England Glass Company in Toledo. In *American Flint* 3, no. 11, September 1912, 45–46.

35. "Boss-Town!"

36. Ibid.

37. "Making Glass To-Day," *Toledo Daily Blade*, August 22, 1888.

38. Scoville, 94.

39. Untitled, *Toledo Daily Blade*, March 1, 1890.

40. Ibid.

41. Untitled, *Fostoria Review*, March 2, 1890.

42. Fauster, 44. The American Flint Glass Workers Union erected a memorial to the workers in St. Mary's Cemetery, Corning, New York.

43. Michael Owens, as quoted by Keene Sumner, "'Don't Try to Carry the Whole World on Your Shoulders!" *American Magazine* 94, July 1922, 127.

44. Ibid.

45. T. W. Rowe, "A History of the American Flint Glass Workers Union," *American Flint* 1, no. 8, June 1910, 21.

46. For example, William S. Walbridge states as such in his unpublished manuscript "The Intimate Lives of Edward Drummond Libbey and Michael Joseph Owens as We Knew Them," 3. In Owens-Illinois, Inc. Company Records, MSS-200, box 20, folder 35. There is no contemporaneous confirmation of this story.

47. Sumner, 127–28.

48. Ibid., 128.

49. Ibid.

50. Pamphlet produced by the Libbey Glass Company for the 1893 fair, titled "Libbey Glass Company's Factory in Full Operation," 2. In Carl U. Fauster Collection, MSS-068, box 1, folder 17.

51. Tudor Jenks, *The Century World's Fair Book for Boys and Girls* (New York: Century Co., 1893), 223.

52. Pamphlet produced by Libbey Glass for the fair, titled "Libbey Glass Company, World's Fair, 1893," 11. In MSS-068, box 1, folder 17.

53. Scoville, 95.

54. Phil Patton, "Sell the Cookstove If Necessary, but Come to the Fair," *Smithsonian* 24, no. 3, 1993, 38.

55. Kate Field, *The Drama of Glass* (Toledo, OH: Libbey Glass Co., 1895).

56. Ibid., 46.

57. Scoville, 96.

58. Rosa B. Lewis, "Florence Scott Libbey," *In Search of Our Past: Women of Northwest Ohio, vol. 2* (Toledo, OH: Women's History Committee of the Women Alive! Coalition, 1990), 18.

59. Scoville, 67.

60. For a discussion of Libbey's brilliant period and the process for producing cut glass, see the unpublished manuscript by Richard A. Schuchert, "Libbey Cut Glass and the Art of the Industry: Background Notes for an Appreciation of Heavy Flint-Glass Cutware from the Libbey Glass Company." In MSS-200, box 22, folder 38.

61. "President's Punch Bowl," *Toledo Blade*, November 30, 1898. As reprinted in Fauster, 80.

62. A photograph depicting John Denman cutting the large punch bowl was featured on the cover of *Scientific American*, April 30, 1904.

63. "Wonderful Cut Glass," *Toledo Blade*, April 11, 1904.

64. As reprinted in Fauster, 259.

65. Scoville, 147. In 2012 dollars, that would be the equivalent of $28.3 million.

66. Ibid. In 2012 dollars, that would be the equivalent of $453 million.

67. Transcript of an oral history interview with John David Biggers, conducted July 18, 1967–April 18, 1968, by Daniel McGinnis, Toledo–Lucas County Public Library, 25. In Libbey-Owens-Ford Glass Company Records, MSS-066, box 7, folder 26.

68. "The Toledo Museum of Art," *Toledo Daily Blade*, April 23, 1901.

69. Ibid.

70. Deed to the 13th and Madison property, dated June 7, 1902. In MSS-200, box 21, folder 44.

71. George W. Stevens, "A Museum of Art—Why?" *Toledo Daily Blade*, January 13, 1912.

72. "George W. Stevens Among Big Directors of Country," *Toledo Daily Blade*, January 13, 1912.

73. Julie A. McMaster, *The Enduring Legacy: A Pictorial History of the Toledo Museum of Art* (Toledo, OH: Toledo Museum of Art, 2001), 14.

74. "Edward D. Libbey Has Given Over $200,000 to Museum," *Toledo Daily Blade*, January 13, 1912.

75. "Enthusiastic Throng Behold Dedication of the Toledo Museum of Art," *Toledo Daily Blade*, January 18, 1912.

76. "Prominent Personages Who Will Take Part in the Dedication of the Toledo Museum of Art," *Toledo Daily Blade*, January 13, 1912.

CHAPTER 3

1. Florence Kelley, "A Boy Destroying Trade: The Glass Bottle Industry of New Jersey, Pennsylvania, Ohio, Indiana, and Illinois," *Charities: A Weekly Review of Local and General Philanthropy* 11, July 4, 1903, 15–19.

2. Ibid., 15.

3. Lee W. Minton, *Flame and Heart: A History of the Glass Bottle Blowers Association of the United States and Canada.* (Washington [DC]: Merkel Press, 1961), 8–9.

4. Warren C. Scoville, *Revolution in Glassmaking* (Cambridge: Harvard University Press, 1948), 30.

5. For further information on automation before Owens, see Edward Meigh, "The Development of the Automatic Glass Bottle Machine: A Story of Some Pioneers," *Glass Technology* 1, February 1960, 26–28.

6. Scoville, 149–50.

7. Articles of incorporation of the Toledo Glass Company, December 16, 1895. In Libbey-Owens-Ford Company Records, MSS-066, box 2, folder 1.

8. "The Lamp Chimney Combine," *National Glass Budget* 15, no. 5, July 11, 1899, 1.

9. Ibid.

10. Meigh describes the workings of the first Owens machine, 30–32.

11. A document titled "Research Notes on Michael J. Owens and the AR Machine," in the Owens-Illinois Company records, quotes engineer Richard LaFrance in his recollections about the development of the bottle machine, recorded in 1959. In Owens-Illinois, Inc. Company Records, MSS-200, box 23, folder 9.

12. Minutes of the Special Meeting of the Stockholders of the Toledo Glass Company, September 16, 1903. In MSS-066, box 2, folder. 1.

13. "The Owens Bottle Machine," *National Glass Budget* 19, no. 14, August 15, 1903, 1.

14. "The Owens Bottle Machine: Its Demonstrated Economy and Efficiency," *National Glass Budget* 19, no. 15, August 29, 1903. 1.

15. Al G. Smith, "The Introduction of the Owens Machine," September 1976, 2. In MSS-200, box 23, folder 9.

16. Check for the first royalty payment to the Owens Bottle Machine Company. In MSS-200, box 45, folder 10.

17. Scoville, 160, contains a chart showing improvements in production capacity for each of the improvements of the Owens Bottle Machine.

18. "Michael Joseph Owens," *Bulletin of the American Ceramic Society—Communications* 17, no. 148, 1939, 150.

19. Minton, 25.

20. Scoville, 203.

21. Ibid., 211.

22. "The Owens Bottle Machine: Its Demonstrated Economy and Efficiency," 1.

23. "The Owens Bottle Machine: Its Relation to the Bottle Industry of the United States," 9. In MSS-200, box 45, folder 15.

24. *Proceedings of the Glass Bottle Blowers Association*, 1906, 33.

25. *Proceedings of the Glass Bottle Blowers Association*, 1908, 26.

26. T. W. Rowe, "History of the American Flint Glass Workers' Union," *American Flint* 1, no. 9, July 1910, 25.

27. T. W. Rowe, "Mechanical Progress and the Owens Machine," *American Flint* 1, no. 6, April 1910, 4.

28. Scoville, 210.

29. Ibid., 205.

30. "Modern Bottles: A Colorful Story," 25. In MSS-200, box 45. folder 9.

31. "Child Labor," *American Flint* 2, no. 12, October 1911, 10.

32. *American Flint* 16, no. 4, February 1925, 3.

33. U.S. Department of Labor Statistics, *Productivity of Labor in the Glass Industry* (Washington, DC: U.S. Department of Labor, 1927), 25.

34. Report to the Stockholders of the Owens Bottle Machine Company, November 9, 1909. In MSS-200, box 45, folder 1.

35. Program for the Fourth Annual Convention of Salesmen, The Owens Bottle Company, Toledo, Ohio, 1923. In MSS-200, box 45, folder 28.

36. Memorandum from Richard LaFrance recalling his work with Michael Owens, September 29, 1959, 8–9. In MSS-200, box 23, folder 9.

37. Ibid, 11.

38. Scoville discusses the management issues regarding Michael Owens in his book on pp. 283–90.

39. Letter from E. D. Libbey to officials, heads of departments, superintendents, office and factory employees, October 16. 1919. In MSS-200, box 21, folder 43.

40. Letter from M. J. Owens to E. D. Libbey, December 5, 1921. In MSS-200, box 21, folder 49.

41. Transcript of an oral history interview with John David Biggers conducted July 18, 1967–April 18, 1968, by Daniel McGinnis, Toledo–Lucas County Public Library, 11. In MSS-066, box 7, folder 26.

42. Memorandum from Richard LaFrance, September 29, 1959, 6.

43. "Random Notes by R. LaFrance on M. J. Owens," August 18, 1958, 6. In MSS-200, box 20, folder 20.

44. Scoville, 308.

45. "The American Society of Mechanical Engineers Designates the Owens 'AR' Bottle Machine as an International Engineering Landmark," 1. In MSS-200, box 6, folder 32.

CHAPTER 4

1. For a review of the early methods of flat glass production, see Arthur E. Fowle, *Flat Glass* (Toledo, OH: Libbey-Owens Sheet Glass Co., 1924); and "History of Plate Glass," *New Albany Daily Ledger*, May 12, 1883.

2. "The Manufacture of Plate Glass in England and the United States," *Scientific American*, September 25, 1869.

3. "Extract from a Diary Kept by Edward Ford during a trip to Europe made with Mr. John Pitcairn in 1893." In Libbey-Owens-Ford Company Records, MSS-066, box 5, folder 27.

4. For a history of the founding of the Edward Ford Plate Glass Company, see William Earl Aiken, *The Roots Grow Deep* (Cleveland, OH: Lezius-Hiles Company, 1957).

5. Ibid., 27.

6. Toledo Glass Company Record Book. In MSS-066, box 1, folder 2.

7. From *A Sketch of the Progress and Development of the Continuous Sheet Glass Process of the Colburn Machine Glass Company*, by Irving W. Colburn, September 2, 1910, as quoted in "Colburn Invents Glass Drawing Machine," *National Glass Budget*, October 27, 1956.

8. As noted in "Invention of a Toledo Man Is Causing a Great Furor in the Glass Industry," *Toledo Daily Blade*, September 2, 1906.

9. Ibid.

10. Ibid.

11. Letter to Stockholders of the Colburn Machine Glass Company, August 16, 1910. In MSS-066, box 54, folder 12.

12. Ibid.

13. Transcript of an oral history interview with John David Biggers conducted July 18, 1967–April 18, 1968, by Daniel McGinnis, Toledo–Lucas County Public Library, 80. In Owens-Illinois, Inc. Company Records, MSS-200, box 7, folder 26.

14. As quoted in Warren C. Scoville, *Revolution in Glassmaking* (Cambridge: Harvard University Press, 1948), 174.

15. George L. Colburn, *The Modern Method of Producing Continuous Sheet Glass* (Quincy, MA: Publisher unknown, 1948). In MSS-066, box 7, folder 29.

16. Fowle, 55.

17. Report to Stockholders by the Libbey-Owens Sheet Glass Board of Directors, December 19, 1919. In MSS-066. box 1, folder 4.

18. Keene Sumner, "Don't Try to Carry the Whole World on Your Shoulders!" *American Magazine* 94, July 1922, 130.

19. Engineering Department Experiment Record Book, Libbey-Owens Sheet Glass, 1921–1924. In MSS-066, box 5, folder 28.

20. Annual report to the stockholders, 1922. In MSS-066, box 4, folder 5.

21. Sumner, 16.

22. Ibid.

23. Ibid., 131.

24. Minutes of the Special Meeting of the Board of Directors of Libbey-Owens Sheet Glass Company, December 29, 1923. In MSS-066, box 1, folder 4.

25. *Toledo Times*, December 28, 1923.

26. *Toledo Blade*, December 28, 1923.

27. "Michael J. Owens," *American Flint* 15, August 1926, 31.

28. William C. Clarke, "Delegates Toledo Welcomes You," *American Flint* 19, July 1929, 3.

29. E. William Fairchild, *Fire & Sand—The History of the Libbey-Owens Sheet Glass Company* (Cleveland, OH: Lezius-Hiles Company, 1960), 113.

30. "Death Claims E. D. Libbey, Glass King," *Toledo Blade*, November 13, 1924.

31. "Libbey's Ideals Spur to Workers," *Toledo News-Bee*, November 13, 1925.

32. Ibid.

33. "A Tribute to Edward Drummond Libbey by his Associates of The Owens Bottle Company." In MSS-200, box 20, folder 5.

34. Minutes of the Stockholders Meeting of the Libbey-Owens Sheet Glass Company, December 9, 1925. In MSS-066, box 1, folder 5.

35. Remarks by James C. Blair, Vice President of Libbey-Owens Sheet Glass Company, Annual Meeting, December 9, 1925. In MSS-066, box 4, folder 5.

36. Annual Report to Stockholders, Libbey-Owens Sheet Glass Company, 1927. In MSS-066, box 4, folder 5.

37. In 2012 dollars, Libbey's estate would be valued at over $272 million.

38. "$10,000,000 Bequeathed in E. D. Libbey Will," *Toledo Blade*, November 19, 1925.

39. "Plate Glass Used by the Automobile Industry, 1924–1929 Inclusive." In MSS-066, box 7, folder 13.

40. "To the Shareholders of Libbey-Owens Glass Company, April 30, 1930." In MSS-066, box 4, folder 6.

41. "To the Shareholders of the Edward Ford Plate Glass Company, May 12, 1930." In MSS-066, box 1, folder 2.

42. Statistics on the demand for glass products as reported in "Gain in Profit Reported by L-O-F," *American Glass Review*, March 3, 1934.

43. Account of the negotiations with the Fisher Body Company, from the transcripts of the oral history interview with John D. Biggers, 36–47.

44. Ibid., 45–46.

45. Ibid., 42.

46. Ibid., 45.

47. "Glass Industry Figure Takes Own Life," *American Glass Review*, August 20, 1932.

48. Minutes of the Board of Directors Meeting, October 25, 1932. In MSS-066, box 1, folder 6.

49. "Gain in Profits Reported by L-O-F."

50. Statistics on employment are from "L-O-F Profits Less on Larger Volume," *American Glass Review*, February 10, 1935.

51. Annual Report of the Libbey-Owens-Ford Company, 1934. In MSS-066, box 4, folder 6.

52. Ibid.

53. "Big Flat Glass Opportunities Ahead," *American Glass Review*, December 15, 1934.

54. Annual Report of the Libbey-Owens-Ford Company, 1937. In MSS-066, box 4, folder 6.

55. "Company Rejects Glass Peace Move," *New York Times*, November 26, 1926.

56. "Trade Body Issues Cease and Desist Order Against Window Glass Distributing Trade," *American Glass Review*, November 6, 1937.

57. "Agreement Ends Libbey-Owens Glass Strike," *Wall Street Journal*, January 28, 1937.

58. Annual Report of Libbey-Owens-Ford Company, reprinted from the Minutes of the Board of Directors, October 11, 1938. In MSS-066, box 4, folder 6.

CHAPTER 5

1. *Annual Report of the Owens Bottle Company for the Year Ending December 31, 1920*; and *Annual Report of the Owens Bottle Company for the Year Ending December 31, 1921*. In Owens-Illinois, Inc. Glass Company Records, MSS-200, box 45, folder 7.

2. All quotes are from Owens Bottle Company annual report, 1921.

3. Address of T. W. Rowe, delivered in Newark, Ohio, March 8, 1925. As published in *American Flint* 15, April 1925, 1.

4. *Annual Report of the Owens Bottle Company for the Year Ending December 31, 1924*. In MSS-200, box 45, folder 7.

5. *Annual Report of the Owens Bottle Company for the Year Ending December 31, 1924*; *Annual Report of the Owens Bottle Company for the Year Ending December 31, 1925*; *Annual Report of the Owens Bottle Company for the Year Ending December 31, 1926*; *Annual Report of the Owens Bottle Company for the Year Ending December 31, 1927*. In MSS-200, box 45, folder 7.

6. As quoted in the unpublished manuscript "The Glassmakers," by William Oursler, 1968, 31–32. In MSS-200, box 18, folder 16.

7. "Soft Drink Prosperity," *Bottles, A Publication of the Illinois Glass Company* 6, September 1919, 14.

8. "Soft Drinks and Bottles," *Bottles, A Publication of the Illinois Glass Company* 7, September 1920, 13.

9. Preston Levis, as quoted in Oursler, 46.

10. H. G. Phillipps, report on acquisition possibility sent to W. H. Boshart, November 13, 1928. In MSS-200, box 46, folder 47.

11. Report on the merger by the Owens Bottle Company, March 28, 1929. In MSS-200, box 46, folder 24.

12. Account of the removal of Boshart in Oursler, 195–99.

13. *Annual Report of the Owens-Illinois Glass Company for the Year Ending December 31, 1929*. In MSS-200, box 1, folder 1.

14. Ibid., 7.

15. *Annual Report of the Owens-Illinois Glass Company for the Year Ending December 31, 1931*, 11. In MSS-200, box 1, folder 1.

16. Owens-Illinois Glass Company annual report, 1929, 8.

17. *Annual Report of the Owens-Illinois Glass Company for the Year Ending December 31, 1930*. In MSS-200, box 1, folder 1.

18. "Average Wage Estimates," *American Flint* 23, April 1935, 26.

19. Oursler, 222–26.

20. James V. Malone, "Sell Bottled Beer: A Reproduction of the Series of Articles Which Appeared in the November, December (1933), and January, February, and March (1934) Issues of *The Brewster and Maltster*." In MSS-200, box 6, folder 20.

21. Ibid., 15.

22. *Annual Report of the Owens-Illinois Glass Company for the Year Ending December 31, 1933*. In MSS-200, box 1, folder 1.

23. *Annual Report of the Owens-Illinois Glass Company for the Year Ending December 31, 1934*. In MSS-200, box 1, folder 1.

24. *Official Book of the Fair: A Century of Progress International Exposition Chicago 1933* (Chicago: Cuneo Press, 1932–33), 17.

25. "Owens-Illinois Glass Containers—1934 Chicago World's Fair." In MSS-200, box 16, folder 30.

26. "Glass—Servant of Man," Owens-Illinois Glass Company, 1931. In MSS-200, box 16, folder 17.

27. C. G. Staelin, unpublished manuscript "Saga of Fiberglas," 4. In Owens Corning Records, MSS-222, box 29, folder 29.

28. "Crystalline Cloth: How the Brittle Cloth is Fashioned," *Pittsburgh Gazette*, November 29, 1880.

29. Staelin, 8.

30. *Annual Report of the Owens-Illinois Glass Company for the Year Ending December 31, 1937*, 22. In MSS-200, box 1, folder 2.

31. Staelin, 26.

32. Ibid., 28–29.

33. "Something to Make of Glass," *Owens Corning Retiree Update*, Fall 2012, 3–4.

34. Faustin J. Solon, "Glass, Servant of Man," *American Flint* 23, February 1936, 1.

35. For the opening of the Newark research center, the company produced a booklet entitled "Alice in Wonderland," which was given to all who attended the dedication. In MSS-222, box 28, folder 25.

36. Owens-Illinois Glass Company annual report for 1937, 20.

37. Staelin, 109.

38. Harold Boeschenstein, draft of memorandum to William Levis, March 12, 1938. In MSS-222, box 29, folder 10.

39. Ibid.

40. Harold Boeschenstein, draft of memorandum to William Levis, April 12, 1938. In MSS-222, box 29, folder 10.

41. Harold Boeschenstein, draft of memorandum to William Levis, May 20, 1938. In MSS-222, box 29, folder 10.

42. Staelin, 112.

43. Harold Boeschenstein, November 1, 1938. In MSS-222, box 29, folder 10.

44. "The Four Basics." In MSS-222, box 62, folder 33.

45. "Owens-Illinois Co. Gets Libbey Glass Business, Assets," *Toledo Blade*, October 18, 1935.

46. For information on the impact of the Nash stemware line on Libbey, see Carl U. Fauster, "Libbey-Nash Stemware Series, 1933," *Hobbies—The Magazine for Collectors*, June 1972, 98J–98M.

47. "Owens-Illinois-Libbey Glass Manufacturing Merger Under Way," *American Flint* 22, May 1938, 20.

48. *Annual Report of the Owens-Illinois Glass Company for the Year Ending December 31, 1935*, 111. In MSS-200, box 1, folder 1.

49. Owens-Illinois Glass Company annual report, 1937, 30.

50. "New Bottle Makes Debut Here," *Toledo Blade*, January 14, 1936.

51. Milton Derber, unpublished manuscript "Collective Bargaining in the Glass Industry," Twentieth Century Fund, New York, 1940, 9. In MSS-200, box 22, folder 13.

52. Ibid., 33.

53. William John Schuck, "Collective Bargaining in the Glass Container Industry, 1917–1937, M.S. thesis, University of Illinois, 1938, 7.

54. United States Department of Labor, "Collective Bargaining in the Glass Industry." As reprinted in *American Flint* 24, July 1936, 3.

55. Derber, 26–27.

CHAPTER 6

1. Morris Dickstein, "From the Thirties to the Sixties: The World's Fair in Its Own Time," in *Remembering the Future: The New York World's Fair From 1939–1964*, Robert Rosenblum, ed. (New York: Rizzoli, 1989), 22.

2. Robert Rydell, *World of Fairs: The Century-of-Progress Expositions* (Chicago: University of Chicago Press, 1993), 215. Rydell speaks to the emphasis of consumerism on the display of "science" at the 1939 fair.

3. *The Miracle of Glass: Its Glorious Past, Its Thrilling Present, Its Miraculous Future as Presented at the New York World's Fair* (Place of publication unknown: Glass Incorporated, 1930), 7.

4. Ibid., 63.

5. *Annual Report of the Owens-Illinois Glass Company for 1938*, 26–27. In Owens-Illinois, Inc. Company Records, MSS-200, box 1, folder 2.

6. *Glass: Building the World of Tomorrow* (Toledo, OH: Owens-Illinois Glass Company, 1939), 12. In MSS-200, box 16, folder 16.

7. Ibid, 21.

8. *American Living and Its Relation to Glass Containers* (Toledo, OH: Owens-Illinois Glass Company, 1941), 19. In MSS-200, box 15, folder 28.

9. Ernie Pyle, *Ernie Pyle on Glass* (Toledo, OH: Libbey Glass, a Division of Owens-Illinois Glass Company, 1945).

10. Transcript of an oral history interview with John David Biggers conducted July 18, 1967–April 18, 1968, by Daniel McGinnis, Toledo–Lucas County Public Library, 102–47. In Libbey-Owens-Ford Company Records, MSS-066, box 7, folder 26–27.

11. "Boeschenstein Asks for Full Effort for War," *Toledo Times*, January 21, 1945.

12. Transcript of Biggers oral history interview, 118–19.

13. Letter from President Franklin Roosevelt to John Biggers, August 29, 1941. In MSS-066, box 79, folder 1.

14. Transcript of Biggers oral history interview, 182.

15. Ibid., 186.

16. Letter from President Franklin Roosevelt to John Biggers, October 30, 1941. In MSS-066, box 79, folder 1.

17. Transcript of Biggers oral history interview, 196.

18. "The Story of Fiberglas—And the Man Who's Putting It Over," *Forbes: The Interpreter of Business*, July 15, 1944, 32.

19. *Annual Report of the Owens-Illinois Glass Company for 1944*, 6. In MSS-200, box 1, folder 2.

20. "The Story of Fiberglas."

21. *Owens-Illinois: People at Work at War* (Toledo, OH: Owens-Illinois Glass Company, 1943), 5. In MSS-200, box 16, folder 34.

22. *Annual Report of the Owens-Illinois Glass Company for 1944*, 23.

23. *Annual Report of the Owens-Illinois Glass Company for 1943*, 5. In MSS-200, box 1, folder 2.

24. "A Special Message to Libbey-Owens-Ford Employes [*sic*]." In MSS-066, box 49, folder 20.

25. Information on labor relations at Libbey-Owens-Ford during the war is detailed in Sean P. Holmes, "Conflict and Cooperation: Labour-Management Relations in the Toledo Flat Glass Industry, 1941–48," M.A. thesis, Bowling Green State University, 1987, 30–47.

26. *Facts About Libbey-Owens-Ford, Submitted to the Navy Department Price Adjustment Board*, 1944. In MSS-066, box 7, folder 28.

27. *Toledo and Lucas County Plan Commissions, What About Our Future?* (Toledo, OH: Lucas County Plan Commission, December 1944), 3.

28. These ads were later compiled in a booklet printed by the *Toledo Blade* to celebrate Toledo's passage of an income tax levy in 1946. The booklet was called *This Is . . . Toledo.*

29. For more on Bel Geddes's career in industrial design, see Dennis P. Doordan, "Toledo Tomorrow," in *The Alliance of Art and Industry: Toledo Designs for a Modern America* (Toledo, OH: Toledo Museum of Art, 2002), 55–65.

30. Letter from Bel Geddes to Henry M. Waite, May 17, 1944. Norman Bel Geddes Papers, Harry Ransom Humanities Research Center, University of Texas.

31. Draft of Bel Geddes's contract with Paul Block Associates, March 20, 1944. Norman Bel Geddes Papers.

32. "Post-War Toledo Depicted in Model," *New York Times*, June 27, 1945.

33. *Toledo Tomorrow, Created by Norman Bel Geddes and Associates* (Toledo, OH: Blade Printing Company, 1945), 1.

34. "Model of Toledo Tomorrow to be On Display July 4," *Toledo Blade*, June 27, 1945, 1.

35. The Metropolitan Planning Committee of the Toledo Chamber of Commerce, *Toward a Master Plan*, 1945.

36. *Toledo Blade* editorial, January 2, 1946.

37. *This Is . . . Toledo*, 31.

38. Transcript of a radio address by John D. Biggers for the Mutual Network, August 30, 1945, 1. In MSS-066, box 79, folder 18.

39. Preston Levis's postwar breakdown is described in the unpublished manuscript "The Glassmakers," by William Oursler, 1969, 298–300. In MSS-200, box 45, folder 7.

40. *Annual Report of the Owens-Illinois Glass Company for 1946*, 7. In MSS-200, box 1, folder 3.

41. *Annual Report of the Owens-Illinois Glass Company for 1948*. In MSS-200, box 1, folder 4.

42. Carl Fauster, *Libbey Glass since 1818: Pictorial History and Collector's Guide* (Toledo, OH: Len Beach Press, 1979), 100.

43. *Annual Report of the Owens-Illinois Glass Company for 1950*, 22–23. In MSS-200, box 1, folder 4.

44. The concept of the solar house was discussed by John Biggers in his radio address of August 30, 1945. The company also produced a promotional kit about the solar house in 1947. In MSS-066, box 18, folder 20.

45. Biggers radio address transcript, 5.

46. Transcript of Biggers oral history, 94–100.

47. Transcript of Biggers oral history, 94–95.

48. *Annual Report of the Libbey-Owens-Ford Company, 1950*, 5. In MSS-066, box 4, folder 8.

49. *Fiberglas: A New Basic Material. Its Development, Properties, Manufacture and Its Uses in War or Peace* (Toledo, OH: Owens-Corning Fiberglas Corporation, 1944). In MSS-222, box 24, folder 24.

50. "Fiberglas Seeks Wide Consumer Market," *Business Week*, December 25, 1948, 48–49.

51. For a summary of these three important legal cases, see Margaret B. W. Graham and Alec T. Shuldner, *Corning and the Craft of Innovation* (New York: Oxford University Press, 2001), 277–90. For further explanation of the impact of the Hartford-Empire case on Owens-Illinois, see Jack K. Paquette, *The Glassmakers Revisited* (Place of publication unknown: Xlibris Corporation, 2010), 62–66.

52. Holmes, 60–77.

53. As quoted in Holmes, 14.

54. "Conciliators to Meet Head of Glass Union," *New York Times*, October 23, 1945.

55. Ibid.

56. Holmes, 94–96.

57. "Owens Centennial Observance Planned in 34 Cities to Honor Unschooled Genius of U.S. Glass Industry," *Toledo Blade*, August 30, 1959.

CHAPTER 7

1. Letter from Wilfred Hibbert, press relations manager at Libbey-Owens-Ford, to H. E. Simpson, professor of glass technology, Alfred University, October 28, 1958. In Libbey-Owens-Ford Company Records, MSS-066, box 14, folder 9.

2. *Annual Report of the Libbey-Owens-Ford Company, 1959*. In MSS-066, box 4, folder 8.

3. Libbey-Owens-Ford press release, July 2, 1951. In MSS-066, box 20, folder 2.

4. "Fiber Glass," *Business Week*, February 26, 1955.

5. Ibid.

6. "Plastics Are Used Boldly in Monsanto's 'House of the Future,'" *Monsanto Magazine*, Summer 1957, 1–8.

7. "F.T.C. See 'Camera Trickery' in Video Ads for Glass Safety," *New York Times*, November 5, 1959.

8. "Pittsburgh Plate Glass Ends Libbey-Owens Grip on GM Business with Major Order," *Wall Street Journal*, April 18, 1961.

9. "Libbey-Owens Sales Net Fell More Than 20% in 1961 from 1960," *Wall Street Journal*, January 30, 1962.

10. "Adventures in Design: A 'Torsion Tower,'" advertising brochure, 1963. In Toledo Mayoral Papers, MSS-061, box 62, folder 1.

11. *Annual Report of the Libbey-Owens-Ford Company, 1968*. In MSS-066, box 4, folder 10.

12. Ibid.

13. *Annual Report of the Libbey-Owens-Ford Company, 1969*. In MSS-066, box 4, folder 10.

14. Evertt Eakin, "LOF to Close Sheet Glass Plant at Shreveport, Louisiana, by June 1," press release, April 7, 1971. In MSS-066, box 23, folder 12.

15. Special reprint, "World's Highest Windows," *Blade Sunday Magazine*, March 28, 1971.

16. Ibid.

17. Letter from Dr. Edward H. Koster, October 20, 1970. In MSS-066, box 50, folder 22.

18. Letter from Arthur Fiedler to Robert Wingerter, May 27, 1974. In MSS-066, box 50, folder 22.

19. As quoted in Frederick C. Fox, *The Rossford Plant of Libbey-Owens-Ford, 1930–1975* (Toledo, OH: Frederick C. Fox, 1982), 138.

20. James C. Haggerty, secretary to the president, White House press release, March 2, 1959. In Owens Corning Records, MSS-222, box 9, folder 27.

21. *Report of the Committee on World Economic Practices*, January 22, 1959. In MSS-222, box 9, folder 27, 1.

22. Ibid.

23. "Fiberglas President Finds USSR's Plants Good but Not Equal to Our Best," *Toledo Blade*, August 16, 1959.

24. Ibid.

25. "Riverview Idea Followed Midnight Stroll," *Toledo Blade*, May 16, 1963.

26. Press release, Office of the Mayor, July 17, 1963. In MSS-061, box 62, folder 3.

27. Ibid.

28. Text of speech by I. M. Pei to the Downtown Toledo Associates, October 3, 1963. In MSS-061, box 62, folder 3.

29. Anonymous letter to Mayor John Potter [1963]. In MSS-061, box 61, folder 22.

30. "O-C Plans Offices Move to Main Riverview Site," *Toledo Blade*, March 22, 1966.

31. James E. Murphy, "Owens-Corning Fiberglas Becomes Major Urban Renewal Spur with Announcement of Plans to Occupy New Office Building," press release, March 22, 1966. In MSS-061, box 62, folder 3.

32. "O-C Plans Offices Move to Riverview Site."

33. "Break Ground for Fiberglas Tower," *Fiberglances*, May 1967.

34. Text of speech delivered by Mayor John Potter to Downtown Toledo Associates, May 1, 1967. In MSS-061, box 61, folder 22.

35. Stephen Stranahan, "The Fiberglas Tower Is Hub of City's Growth," *Tower News*, 1, no. 2, June 1968.

36. Mary Kincaid, "The Art Collection at Owens-Corning Isn't Investment: It's Inspiring," *American Glass Review*, March 1981.

37. "New Uses of Fibrous Glass," *Business Week*, April 11, 1953.

38. *Annual Report of Owens-Corning Fiberglas Corporation, 1957*. In MSS-222, box 2, folder 1.

39. "Harold Boeschenstein Retires as OCF Chief Officer; General Norstad Elected Chairman," *Fiberglances*, October 1967.

40. *Annual Report of Owens-Corning Fiberglas Corporation, 1965*. In MSS-222, box 1, folder 3.

41. *The United States Senate Report of Proceedings—Hearing before the Committee on Aeronautical and Space Sciences, Apollo Accident, February 27, 1967*. In MSS-222, box 2, folder 6.

42. Letter from Geoffrey M. Kalmus to John H. Thomas, August 27, 1969. In MSS-222, box 2, folder 8.

43. Conference Kit on Fire Safety, 1970. In MSS-222, box 2, folder 20.

44. "Drive to Get Results Spurred Boeschenstein," *Toledo Blade*, October 23, 1972.

45. "Fiber Glass Growth Based on Its Ability to Solve Problems," *American Glass Review*, July 1976, 29.

46. *Gateway to Mecca: The Story of the Hajj Terminal* (Toledo, OH: Owens-Corning Fiberglas, 1978), 7. In MSS-222, box 3, folder 28.

47. Ibid., 15.

48. Ibid, 27.

49. "Owens-Corning Chief Lists Four Future Threats to Firm," *Toledo Blade*, April 24, 1979.

50. "They're Making Big Glass in Texas," *American Glass Review*, June 1979, 7.

51. Transcript of an oral history interview with William Boeschenstein, recorded September 5, 2001. In MSS-222, box 33, folder 19.

52. *Annual Report of Owen-Illinois Glass Company, 1954*. In Owens-Illinois, Inc. Company Records, MSS-200, box 1, folder 5.

53. *Annual Report of the Owens-Illinois Glass Company, 1960*. In MSS-200, box 1, folder 5.

54. *Annual Report of the Owens-Illinois Glass Company, 1956*. In MSS-200, box 1, folder 5.

55. "Firm Goes Pining in the Bahamas," *Toledo Blade Sunday Magazine*, July 24, 1960, 7.

56. "Report on the Proposed Cane Sugar Operations on Great Abaco Island," 1966. In MSS-200, box 2, folder 33.

57. Ibid.

58. Ibid.

59. *Annual Report of the Owens-Illinois Inc., 1970*. In MSS-200, box 1, folder 5.

60. Resolution of the Owens-Illinois Board of Directors, December 19, 1962. In MSS-200, box 1, folder 5.

61. R. H. Mulford, *The Integrated Work Force: Where Are We Now?* (New York: American Management Association, 1962), 2.

62. *Annual Report of the Owens-Illinois Inc., 1970.* In MSS-200, box 1, folder 7.

63. *Annual Report of the Owens-Illinois Inc., 1971.* In MSS-200, box 1, folder 7.

64. *Annual Report of the Owens-Illinois Inc., 1970.*

65. Jerry Arkebauer, Owens-Illinois News Bureau, August 15, 1977. In MSS-200, box 11, folder 27.

66. *Annual Report of the Owens-Illinois Inc., 1974.* In MSS-200, box 1, folder 7.

67. Ibid.

68. *Annual Report of the Owens-Illinois Inc., 1979.* In MSS-200, box 1, folder 8.

69. *Annual Report of the Libbey-Owens-Ford Company, 1974.* In MSS-066, box 4, folder 13.

70. Ibid.

71. *Annual Report of the Libbey-Owens-Ford Company, 1976.* In MSS-066, box 4, folder 12.

72. Ibid.

73. Ibid.

74. Ibid.

75. Exhibit Center files, 1974. In MSS-222, box 6, folder 6.

76. "The Fiberglass Boom: Will It Last?" *Forbes*, July 24, 1978, 50.

77. Anthony J. Paris, "Owens-Corning Soars on the Fiberglass Boom," *New York Times*, June 22, 1978.

78. *Annual Report of the Owens-Illinois Inc., 1974.* In MSS-200, box 1, folder 7.

79. *Annual Report of the Owens-Illinois Inc., 1981.* In MSS-200, box 1, folder 8.

80. *Annual Report of the Owens-Illinois Inc., 1979.* In MSS-200, box 1, folder 8.

81. "Kaylo Joins the Fiberglas Family," *Fiberglances*, July–August, 1958, 4.

82. *Annual Report of Owens-Corning Fiberglas Corporation, 1981.* In MSS-222, box 1, folder 5.

CHAPTER 8

Note: Portions of this chapter were previously published in *Northwest Ohio History* 77, no. 1, Fall 2009. Used with permission.

1. Owens-Illinois News Bureau, "Chronology—History of Owens-Illinois, Inc. in Downtown Toledo," for 1969. In Owens-Illinois, Inc. Company Records, MSS-200, box 7, folder 28.

2. Owens-Illinois News Bureau, "Chronology—History of Owens-Illinois, Inc. in Downtown Toledo," for 1971. In MSS-200, box 7, folder, 28.

3. In 1973, the average unemployment rate for Lucas County was 4.7 percent. During 1974, the average climbed to 5.8 percent. In 1975, the rate hit 12.3 percent in February before declining to 8.2 percent in December. *Ohio Labor Force Estimates by County, by Month 1970–1977* (Columbus, OH: Bureau of Employment Services, Division of Research and Statistics, 1977).

4. Convention Committee minutes. In MSS-027, Downtown Toledo Associates Records, box 1, folder 6.

5. Ibid.

6. "Charter Change Wins; Project Delays Expected," *Toledo Blade*, March 21, 1973, 1.

7. "Civic Setback," *Toledo Blade*, March 21, 1973, 16.

8. ULI—The Urban Land Institute, "An Evolution of the Redevelopment Potential of Downtown Toledo for the Toledo CBD Study Group," 1973. In MS-47, Harry Kessler Papers, box 29, folder 5, Center for Archival Collections, Bowling Green State University.

9. Ibid., 19.

10. Ibid., 23.

11. Randy Sparks, as interviewed by Folk Alley Extras, available at http://www.podcastdirectory.com/podshows/40399. Accessed on February 12, 2008.

12. O-I News Bureau, "Chronology—History of Owens-Illinois in Downtown Toledo," for 1974. In MSS-200. box 17, folder 28.

13. EPD Environmental Planning and Design, *Toledo Looks to the River* (Pittsburgh: EDP, 1975).

14. "GTC Activities since Inception," Stephen Stranahan Papers. Private collection.

15. O-I News Bureau, "Chronology—History of Owens-Illinois in Downtown Toledo," for 1975. In MSS-200, box 17, folder 28.

16. Letter and attachment by Stephen Stranahan to the University of Toledo Board of Trustees, December 17, 1975. Stephen Stranahan Papers.

17. In MS-47, box 3, folder 3.

18. Ibid., box 30, folder 3.

19. O-I News Bureau, "Chronology—History of Owens-Illinois in Downtown Toledo," for 1976. In MSS-200, box 17, folder 29.

20. Ibid., 4–5.

21. O-I News Bureau, "Chronology—History of Owens-Illinois in Downtown Toledo," for 1977. In MSS-200, box 17, folder 29.

22. Letter from Rohrbacher to Kessler. In MS-47, box 30, folder 2.

23. Ibid.

24. Letter from Stranahan to David Drury, January 19, 1978. Stephen Stranahan Papers.

25. O-I News Bureau, "Chronology—History of Owens-Illinois in Downtown Toledo," for 1979. In MSS-200, box 17, folder 32.

26. O-I News Bureau, "Chronology—History of Owens-Illinois in Downtown Toledo," for 1971. In MSS-200, box 17, folder 28.

27. O-I News Bureau, "Chronology—History of Owens-Illinois in Downtown Toledo," for 1976. In MSS-200, box 17, folder 29.

28. Minutes of the TDC, July 18, 1977. In MSS-200, box 2, folder 42.

29. Tape recording, news conference on Feasibility Study Completion, 1977. In MSS-200, box 49, folders 18–19.

30. Speech by Dodd to executive committee of the Owens-Illinois Board of Directors, October 5, 1977. In MSS-200, box 2, folder 40.

31. Memorandum from Jerry Arkebauer to Sam Allen, November 17, 1978. In MSS-200, box 3, folder 36.

32. *Toledo Blade*, January 17, 1979, 20.

33. Max Abrahamovitz, talk to the American Institute of Architects, May 10, 1979. In MSS-200, box 2, folder 48.

34. Ted Ligibel, as quoted in "Decay Foils Bid to Save Façade of Former Bank," *Toledo Blade*, April 12, 1979.

35. "Enthusiasm of Speakers, Observers Mark 'The Start of Something Big,'" *Toledo Blade*, May 23, 1979.

36. Letter from Colin Smith, correspondence editor, Guinness Superlatives Limited, July 24, 1979. In MSS-200, box 2, folder 48.

37. Letter from Leslie Barr to Willard Webb III, March 12, 1975. In MS-47, box 30, folder 1.

38. "Bringing Urban Glitter to Smaller Cities," *Business Week*, July 23, 1984, 140.

39. O-I News Bureau, "Chronology—History of Owens-Illinois in Downtown Toledo," for 1979. In MSS-200, box 17, folder 32.

40. Draft of press release by John Hoff, O-I News Bureau, to J. N. Graham and D. J. Toth, March 19, 1980. In MSS-200, box 3, folder 3.

41. Minoru Yamasaki, "River View Should Be Preserved," *Toledo Blade*, October 5, 1980.

42. "Walling Off the River," *Toledo Blade*, April 20, 1980.

43. Intra-Company Memoranda from John Hoff to Jack Paquette, May 28, 1980. In MSS-200, box 3, folder 5.

44. *Toledo Blade*, July 28, 1980, 6.

45. Ibid.

46. "The Downtown Dream," *Toledo Blade*, November 23, 1981.

47. William Brower, "Portside Good Omen for Toledo," *Toledo Blade*, May 19, 1984, 8.

48. "Bringing Urban Glitter to Smaller Cities," *Business Week*, July 23, 1984, 140.

49. Anonymous city leader, as quoted in "Is SeaGate III the Next Step?" *Toledo Blade*, March 17, 1987, section B, p. 1. The story, by *Blade* writer Dave Murray, outlines the complex financing scheme that backed downtown development.

50. "SeaGate Center Readies for March 27 Dedication," *Toledo Blade*, March 5, 1987.

51. "Downsizing Old Hat for Some Toledo Firms," *Toledo Blade*, Mary 22, 1990,

52. "First Fed's Dismemberment," *Toledo Blade*, October 2, 1991, 6.

53. "It Was a Year of Change for Toledo," *Toledo Blade*, January 1, 1990, 1.

54. "Searching for a Center," *Toledo Blade*, January 1, 1990, 10.

CHAPTER 9

1. Homer Brickey, "Owens-Illinois leaving Toledo for Perrysburg," *Toledo Blade*, May 6, 2003.

2. Jon Chavez, "3 firms Consider Downtown Exit," *Toledo Blade*, May 4, 2003.

3. Gary T. Pakulski and Mary-Beth McLaughlin, "Bankrupt OC May Move from Headquarters Building," *Toledo Blade*, March 22, 2003.

4. Bruce Vernyi, "Glass Operators Sale to Give LOF Different Identity," *Toledo Blade*, March 13, 1986.

5. "Extract from a Diary Kept by Edward Ford during trip to Europe made with Mr. John Pitcairn in 1893." In Libbey-Owens-Ford Company Records, MSS-066, box 5, folder 27.

6. Homer Brickey, "LOF Glass Intends to Adopt Name of Libbey-Owens-Ford," *Toledo Blade*, July 9, 1986.

7. Homer Brickey, "New Name Proposed for LOF is Trinova—'3 Bright Stars,'" *Toledo Blade*, July 8, 1986.

8. "Sale of LOF to Pilkington Final," *Toledo Blade*, April 29, 1986.

9. "Pilkington Says Legal Issues Gave It an Edge," *Toledo Blade*, May 6, 2003.

10. Ibid.

11. Ibid.

12. Owen-Corning Fiberglas press release, August 6, 1986. In Owens Corning Records, MSS-222, box 9, folder 16.

13. "Owens-Corning Gets Wickes Bid of $70 a Share," *Wall Street Journal*, August 7, 1986.

14. "Chairman of Owens Fights Hostile Bid," *New York Times*, August 14, 1986.

15. Owens-Corning press release, August 6, 1986.

16. Owens-Corning press release, "Owens-Corning Fiberglas Corporation Rejects Offer by Wickes Companies, Inc.," August 23, 1986. In MSS-222, box 9, folder 16.

17. Transcript of an oral history interview with William Boeschenstein, recorded September 5, 2001. In MSS-222, box 33, folder 19.

18. "Bid to Prevent Take-Over Will Cost Owens-Corning at Least $750,000," *Toledo Blade*, August 27, 1986.

19. Bruce Vernyi, "Wickes Halts Take-Over Bid for Owens-Corning, Nets $9.4 Million Profits," *Toledo Blade*, August 30, 1986.

20. Ibid.

21. "It's Full Steam Ahead at the New Owens-Corning," *Toledo Blade*, October 5, 1986.

22. Ibid.

23. Numbers of layoffs at each facility are from an October 15, 1986, letter issued to all Owens-Corning personnel. In MSS-222, box 9, folder 17.

24. Homer Brickey, "Boeschenstein Says OCF Will Cut 650 Jobs from Its Toledo Headquarters," *Toledo Blade*, October 15, 1986.

25. Homer Brickey, "Job Loss Paranoia Grows on Fear at OCF," *Toledo Blade*, September 23, 1986.

26. Figures on cost of restructuring from Pat Green, "OCF Stockholders OK Recapitalization," *Toledo Blade*, November 5, 1986.

27. Michael D. Towle, "O-I to Sell Forest Products Division for $1.5 Billion," *Toledo Blade*, July 16, 1987.

28. Laurie Krauth," O-I Subsidiary Given OK for Nursing Home," *Toledo Blade*, July 23, 1987.

29. Michael D. Towle, "O-I Seeking 2nd-Biggest Glass-Container Maker," *Toledo Blade*, September 7, 1987.

30. Michael D. Towle, "FTC Appeal Rejected; Way Cleared for O-I to Acquire Brockway," *Toledo Blade*, February 27, 1988.

31. Pat Welsh, "Mergers, Takeovers, and Leveraged Buy-Outs—Who Pays?" *Metropolitan*, October 1987, 35.

32. Ibid.

33. Ibid., 38.

34. Michael D. Towle, "Life after O-C Started the Day the Boss Called," *Toledo Blade*, November 2, 1986.

35. "The End of an Automotive Era," *Toledo Blade*, October 20, 1991.

36. Michael D. Sallah and Mark Rollenhagen, "The Wait: Jobless Arrive Early to Stand in Long Lines at Benefits Office," *Toledo Blade*, January 5, 1990.

37. Editorial, "A Deal Is Still a Deal," *Toledo Blade*, April 5, 1991.

38. Libbey, Inc. annual report, 1994. In Jack Paquette Collection on Northwest Ohio's Glass Industry, MSS-169, box 2, folder 17.

39. For a history of early reports of Kaylo's health issues, see Paul Brodeur, *Outrageous Misconduct: The Asbestos Industry on Trial* (New York: Pantheon Books, 1985), 148–54.

40. Ibid., 150.

41. As quoted in ibid., 150.

42. Ibid.

43. As quoted in ibid., 151.

44. "Owens-Corning Fiberglas Purchases Kaylo Division of Owens-Illinois," *Fiberglances*, June 1958. In MSS-222, box 27, folder 41.

45. Cynthia F. Mitchell and Paul M. Barrett, "Trial and Error: Novel Effort to Settle Asbestos Claims Fails as Lawsuits Multiply," *Wall Street Journal*, June 7, 1988.

46. "Asbestos Suit Settled on Eve of Trial," *Toledo Blade*, October 2, 1992.

47. Owens Corning Financial Reorganization—Asbestos Chronology, available at http://www.owenscorning.com/finre/asbestos.html. Accessed on December 17, 2012.

48. Ibid.

49. Gary T. Pakulski, "Questions Outnumber Answers on Fallout from OC Headquarters," *Toledo Blade*, March 25, 2003.

50. Gary T. Pakulski, "Burned by Asbestos," *Toledo Blade*, April 2, 1995.

51. "A History of Owens-Corning," *Toledo Blade*, October 6, 2000.

52. Owens Corning Financial Reorganization website.

53. Gary T. Pakulski, "O-C Will Pay $1.2 Billion to Settle Most Asbestos Cases," *Toledo Blade*, December 16, 1998.

54. Gary T. Pakulski, "No Bankruptcy, OC Vows, but Can It Make Enough to Pay Liability Claims?" *Toledo Blade*, July 2, 2000.

55. Gary T. Pakulski, "'Overwhelming' Liability on Asbestos Forces Action," *Toledo Blade*, October 6, 2000.

56. Ibid.

57. Roberta DeBoer, "Relief Plans May Provide Too Much Insulation: Proposed Aid for Asbestos Firms Elicits Mixed Feelings," *Toledo Blade*, October 22, 2000.

58. "OC Develops Bonus Plan to Keep Key Staff," *Toledo Blade*, January 18, 2001.

59. "OC Loses $2.8 Billion in '02; Charge for Asbestos Cited," *Toledo Blade*, February 11, 2003.

60. Gary T. Pakulski, "Missteps and Dissensions: OC Bankruptcy Documents Reveal Rifts in Firm," *Toledo Blade*, December 8, 2002.

61. Gary T. Pakulski, "OC Submits Plan to Exit Bankruptcy," *Toledo Blade*, January 18, 2003.

62. Gary T. Pakulski and Mary-Beth McLaughlin, "Bankrupt OC May Move out of Headquarters Building," *Toledo Blade*, March 22, 2003.

63. Editorial, "One OC Hurdle Cleared," *Toledo Blade*, April 14, 2003.

64. Tahree Lane, "O-I Asbestos Claims Approved by Court," *Toledo Blade*, December 22, 1995.

65. Mary-Beth McLaughlin, "O-I Chief Dismisses Bankruptcy Fear," *Toledo Blade*, October 7, 2000.

66. Gary T. Pakulski, "Accord Near on O-I Loan," *Toledo Blade*, April 17, 2001.

67. Editorial, "Ed Dodd's legacy," *Toledo Blade*, January 7, 2001.

68. Tom Troy, "Owens-Illinois' Future Downtown Is Unclear," *Toledo Blade*, January 29, 2005.

69. Ibid.

70. Ibid.

71. Homer Brickey, "O-I Chief Says Odds Favor Move; Perrysburg Is Preferred, but Toledo Still Has a Chance," *Toledo Blade*, February 4, 2005.

72. Homer Brickey, "Work Force Enamored with Move," *Toledo Blade*, May 7, 2005.

73. Home Brickey, "Owens-Illinois Leaving Toledo for Perrysburg," *Toledo Blade*, May 6, 2005.

74. Homer Brickey, "Fifth Third Bank Verifies Its Interest in One SeaGate," *Toledo Blade*, February 10, 2006.

75. Ibid.

76. Homer Brickey, "LOF Glass Intends to Adopt Name of Libbey-Owens-Ford."

77. As quoted in "British Firm Sheds L-O-F Identity," *Toledo Blade*, May 27, 2000.

78. "One Pilkington," *Toledo Blade*, May 27, 2000.

79. Julie M. McKinnon, "Rossford L-O-F plant Challenged by Widespread Turnover," *Toledo Blade*, March 28, 2000.

80. Editorial, "A Symbolic Passing," *Toledo Blade*, June 1, 2000.

81. Tom Troy, "Developer Set to Buy Pilkington Building," *Toledo Blade*, August 13, 2002.

82. Ibid.

83. Tom Troy, "Hyland, Pilkington to Stay Downtown—City's Incentive Plan Seals Deal," *Toledo Blade*, June 14, 2003.

84. Ibid.

85. Gary T. Pakulski, "Flint Glass Union Delegates Vote on Merger into Steelworkers," *Toledo Blade*, June 3, 2003.

86. Edward Drummond Libbey and Florence Scott Libbey estate distribution summary, 1928–1975. In Owens-Illinois, Inc. Company Records, MSS-200, box 21, folder 47.

87. Ibid.

88. For more information on the beginnings of the studio glass movement, see Jutta-Annette Page, *The Art of Glass* (Toledo, OH: Toledo Museum of Art, 2006), 83–88.

89. Heather Denniss, "A Far Reaching Project: Museum Taps Resources from around the World," special supplement to the *Toledo Blade*, August 20, 2006.

90. James T. Areddy, "In Toledo, the 'Glass City,' New Label: Made in China," *Wall Street Journal*, August 29, 2010.

91. Denniss.

92. Homer Brickey, "Pioneer's House to Showcase Glass City—Nonprofit to Buy Libbey Mansion," *Toledo Blade*, September 25, 2007.

93. Areddy.

94. "Inventor Became Philanthropist," *Toledo Blade*, August 26, 2003.

95. "Glasstech Builds on Pioneering Ideas," *Glasstech World: Special Edition*, December 4, 2008, 2.

96. Ibid., 3.

97. Brook Sopelsa, "Toledo Tries to Overcome Rust-Belt Image, Become Green Jobs Metropolis," available at http://www.cnbc.com/id/31385869. Accessed on April 30, 2013.

98. As quoted in Sopelsa.

99. Peter C. Glover, "Solar Eclipsed?" *Energy Tribune*, October 24, 2012, available at http://www.energytribune.com/63932.solar-eclipsed-2. Accessed April 26, 2013. Glover reviews the recent crisis in the solar energy business, which has produced financial difficulties for many firms.

100. Jim Siegel, "3 Toledo-Area Firms Fail to Make Loan Payments. State Officials Try to Balance Aid, Proper Use of Funds," *Columbus Dispatch*, April 15, 2012.

101. Ibid.

102. Kris Turner, "Execs: Outlook Sunny at Xunlight Solar Firm," *Toledo Blade*, May 2, 2013.

103. "In 1st year, Transition Smooth at Glass Firm," *Toledo Blade*, June 19, 2007.

104. "Venezuelan Takeover Deals O-I Loss for '10," *Toledo Blade*, January 27, 2011.

105. Ignazio Messina, "City, OC Reach Deal to Keep Firm until 2030," *Toledo Blade*, October 19, 2013.

Bibliography

UNPUBLISHED SOURCES

American Flint Glass Workers Union Records, West Virginia Museum of American Glass, Weston, West Virginia.

Carl U. Fauster Collection, MSS-068, Ward M. Canaday Center for Special Collections, University of Toledo.

Dominick Labino Collection, MSS-223, Ward M. Canaday Center for Special Collections, University of Toledo.

Downtown Toledo Associates Records, MSS-027, Ward M. Canaday Center for Special Collections, University of Toledo.

Harry J. Durholt Papers, MSS-063, Ward M. Canaday Center for Special Collections, University of Toledo.

Harry Kessler Papers, MS-47, Center for Archival Collections, Bowling Green State University.

Jack Paquette Collection on Northwest Ohio's Glass Industry, MSS-169, Ward M. Canaday Center for Special Collections, University of Toledo.

John L. Lewis Papers, MSS-201, Ward M. Canaday Center for Special Collections, University of Toledo.

Libbey-Owens-Ford Glass Company Records, MSS-066, Ward M. Canaday Center for Special Collections, University of Toledo.

Norman Bel Geddes Papers, Harry Ransom Humanities Research Center, University of Texas.

Owens Corning Records, MSS-222, Ward M. Canaday Center for Special Collections, University of Toledo.

Owens-Illinois, Inc. Company Records, MSS-200, Ward M. Canaday Center for Special Collections, University of Toledo.

Stephen Stranahan Papers, private collection.

Toledo Mayoral Papers, MSS-061, Ward M. Canaday Center for Special Collections, University of Toledo.

Toledo Museum of Art Archives, Toledo Museum of Art.

MONOGRAPH SOURCES

Aiken, William Earl. *The Roots Grow Deep*. Cleveland, OH: Lezius-Hiles Company, 1957.

Brodeur, Paul. *Outrageous Misconduct: The Asbestos Industry on Trial*. New York: Pantheon Books, 1985.

Colburn, George L. *The Modern Method of Producing Continuous Sheet Glass*. Quincy, MA: Publisher unknown, 1948.

Doordan, Dennis P. *The Alliance of Art and Industry: Toledo Designs for a Modern America*. Toledo, OH: Toledo Museum of Art, 2002.

Downes, Randolph C. *Industrial Beginnings: Lucas County Historical Series*. Vol. 4. Toledo, OH: Toledo Printing Company, 1954.

Drake, Daniel. *A Systematic Treatise, Historical, Etiological, and Practical, on the Principal Diseases of the Interior Valley of North America*. Cincinnati, OH: Winthrop B. Smith & Co., 1850.

Faber, Don. *The Toledo War: The First Michigan-Ohio Rivalry*. Ann Arbor: University of Michigan Press, 2008.

Fairfield, E. William. *Fire & Sand: The History of the Libbey-Owens Sheet Glass Company*. Cleveland, OH: Lezius-Hiles Company, 1960.

Fauster, Carl U. *Libbey Glass since 1818: Pictorial History and Collectors Guide*. Toledo, OH: Len Beach Press, 1979.

Field, Kate. *The Drama of Glass*. Toledo, OH: Libbey Glass Co., 1895.

Fowle, Arthur E. *Flat Glass*. Toledo, OH: Libbey-Owens Sheet Glass Co., 1924.

Fox, Frederick C. *The Rossford Plant of Libbey-Owens-Ford, 1930–1975*. Privately published by author, 1982.

Graham, Margaret B. W. and Alec T. Shuldiner. *Corning and the Craft of Innovation*. New York: Oxford University Press, 2001.

Hartung, Walter, ed. *History of Medical Practice in Toledo and the Maumee Valley Area, 1600–1990*. Place and date of publication unknown.

Holmes, Sean P. "Conflict and Cooperation: Labour-Management Relations in the Toledo Flat Glass Industry, 1941–48." M.A. thesis, Bowling Green State University, 1987.

Jenks, Tudor. *The Century World's Fair Book for Boys and Girls*. New York: Century Co., [1893].

McMaster, Julie. *The Enduring Legacy: A Pictorial History of the Toledo Museum of Art*. Toledo, OH: Toledo Museum of Art, 2001.

Minton, Lee W. *Flame and Heart: A History of the Glass Bottle Blowers Association of the United States and Canada*. Washington [DC]: Merkel Press, 1961.

Miracle of Glass: Its Glorious Past, Its Thrilling Present, Its Miraculous Future as Presented at the New York World's Fair. Place of publication unknown: Glass Incorporated, 1939.

Monro, William L. *Window Glass in the Making: An Art, a Craft, a Business*. Pittsburgh: American Window Glass Company, 1926.

Official Book of the Fair: A Century of Progress International Exposition Chicago 1933. Chicago: Cuneo Press, Inc., 1932–33.

Page, Jutta-Annette. *The Art of Glass: Toledo Museum of Art.* Toledo, OH: Toledo Museum of Art, 2006.

Paquette, Jack K. *The Glassmakers: A History of Owens-Illinois, Incorporated.* Toledo, OH: Trumpeting Angel Press, 1994.

Paquette, Jack K. *The Glassmakers Revisited.* Place of publication unknown: Xlibris Corporation, 2010.

Pittsburgh Plate Glass Company. *Glass: History, Manufacture and Its Universal Application.* Pittsburgh: Pittsburgh Plate Glass Company, 1923.

Porter, Tana Mosier. *Toledo Profile: A Sesquicentennial History.* Toledo, OH: Toledo Sesquicentennial Commission, 1987.

Pyle, Ernie. *Ernie Pyle on Glass.* Toledo, OH: Libbey Glass, A Division of Owens-Illinois Glass Company, 1945.

Rydell, Robert. *World of Fairs: The Century-of-Progress Expositions.* Chicago: University of Chicago Press, 1993.

Schuck, William John. "Collective Bargaining in the Glass Container Industry, 1917–1937." M.S. thesis, University of Illinois, 1938.

Scribner, Harvey, ed. *Memoirs of Lucas County and the City of Toledo: From the Earliest Historical Times Down to the Present, Including a Genealogical and Biographical Record of Representative Families.* Madison, WI: Western Historical Association, 1910.

Scott, Jesup W. *A Presentation of Causes Tending to Fix the Position of the Future Great City of the World in the Central Plain of North America.* Toledo, OH: Blade Printing & Paper Co., 1937.

Scoville, Warren C. *Revolution in Glassmaking: Entrepreneurship and Technological Change in the American Industry, 1880–1920.* Cambridge: Harvard University Press, 1948.

Waggoner, Clark, ed. *History of the City of Toledo and Lucas County, Ohio.* New York: Munsell & Company, Publishers, 1888.

Watkins, Lura Woodside. *Cambridge Glass, 1818 to 1888.* New York: Bramhall House, 1930.

Winter, Nevin O. *A History of Northwest Ohio.* Chicago: Lewis Publishing Company, 1917.

Women's History Committee of the Women Alive! Coalition. *In Search of Our Past: Women of Northwest Ohio. Vol. 2.* Toledo, OH: Womens's History Committee of the Women Alive! Coalition, 1990.

Index

Page numbers in italics refer to illustrations.

Printed and bound by CPI Group (UK) Ltd, Croydon, CR0 4YY
16/04/2025
14658541-0001